The Political Dynamics of School Choice

The Political Dynamics of School Choice

Negotiating Contested Terrain

Lance D. Fusarelli

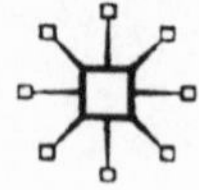

THE POLITICAL DYNAMICS OF SCHOOL CHOICE

First published 2003 by
PALGRAVE MACMILLAN™
175 Fifth Avenue, New York, N.Y. 10010
and Houndmills, Basingstoke, Hampshire, England RG21 6XS.
Companies and representatives throughout the world.

PALGRAVE MACMILLAN is the global academic imprint of the Palgrave Macmillan division of St. Martin's Press, LLC and of Palgrave Macmillan Ltd. Macmillan® is a registered trademark in the United States, United Kingdom and other countries. Palgrave is a registered trademark in the European Union and other countries.

ISBN 0–312–23753–7 hardback—1–4039–6047–X paperback

Library of Congress Cataloging-in-Publication Data
Fusarelli, Lance D. (Lance Darin), 1966–
The political dynamics of school choice: negotiating contested terrain/Lance D. Fusarelli.
p. cm.
Includes bibliographical references and index.
ISBN 0–312–23753–7 (hbk)—ISBN 1–40396–047–X (pbk.)
1. School choice—United States. 2. Education and state—United States. I. Title.

LB1027.9.F87 2003
379.1'11'0973—dc21 2002193056

A catalogue record for this book is available from the British Library.

Design by Newgen Imaging Systems (P) Ltd., Chennai, India.

First edition: May, 2003
10 9 8 7 6 5 4 3 2 1

Printed in the United States of America.

To my mother, whose great desire was to write a book.
That her wish may be fulfilled through her son.
I miss you

Table of Contents

Acknowledgments

This book could not have been written without the support of several individuals and organizations. In particular, I would like to thank my dear friend and colleague, Bruce Cooper, who first encouraged me to write this book. His thoughtful review and critique of the book improved the final product. I was able to complete this task due in no small measure to his constant pestering. Speaking of which, special thanks go to my mentor, friend, and surrogate father, Jay Scribner. I hope I am as good a mentor to my students as he was to me.

I am indebted to my editor, Michael Flamini, and assistant editor, Amanda Johnson, at Palgrave Macmillan, for their great patience when I missed repeated deadlines, able assistance in revising my prose and making it intelligible, and encouragement and support. I would like also to thank the reference staffs of the Cleveland, New York, and Pittsburgh Public Libraries as well as those at Fordham University and the University of Texas at Austin for their invaluable assistance in retrieving documents and for allowing me to monopolize their copy machines for extended periods of time.

Most of all, I would like to thank the many individuals whom I interviewed in the course of this project. These dedicated, committed individuals took time from their always hectic schedules to answer my queries and shed light on the dynamics of state politics and the school choice movement. Of course, despite all their assistance, any errors of fact or logic are mine alone.

List of Tables

Chapter 1

School Choice and the Politics of Reform

Few educational reform proposals generate as much vitrolic rhetoric or outright hostility as proposals for school choice—be they charter schools, tuition tax credits, privatization, or vouchers.[1] Political battles over school choice are viewed as life-or-death struggles for the soul of American education. In some respects, this view is correct and not overly melodramatic, since school choice in its purest form—a full-blown voucher plan, for example—threatens to change the way schools are operated, funded, and governed. Choice proposals represent a challenge (and a threat) to nearly every constituency in the traditional education establishment—administrators, teachers, their unions, professional and nonprofessional staff, school boards, even state education agencies and schools of education. For example, charter school laws in several states provide waivers from "laws and regulations that teacher unions have spent generations setting in place to enhance their members' interests" such as collective bargaining and teacher certification.[2] Voucher plans threaten the hegemony and power of the Education Establishment because they reallocate resources from the public to the private sector. Since the potential impact of school choice proposals is so dramatic, the stakes are raised to an unprecedented level. Accordingly, studying the political dynamics of school choice is important, since it exposes the major value differences in society and highlights politics and policymaking in its rawest and most interesting form.[3]

This book explores the politics of school choice at the state level—where the real political action occurs. Although the principal focus of the text is on the political battles over charter schools and vouchers, it is not simply yet another book about school choice. The purpose is to treat the voucher and charter school "wars" as illustrative case studies of the politics of educational change—cases that illuminate much of the oft-concealed politics of state-level policymaking. Throughout this book, the author demonstrates how the tensions, constraints, and political battles over choice initiatives illustrate the richness and complexity of state policymaking. Drawing from John Kingdon's pioneering study of agendas, alternatives, and public policy,[4] I explore how political culture and language, interest groups, institutions, and organizational (policy) learning come together to shape the politics, process, and progress of school choice and state education reform.

As a study of the dynamics of state policymaking, the politics of school choice has implications for other areas of state-level policy activism such as the policy arenas of health, transportation, justice/corrections, and the environment, as well as for other issues within education itself. It is reasonable to ask how a book about the politics of school choice could possibly be relevant to understanding other policy areas. The answer is obvious. The basic institutional structure of the state remains the same—the legislative committee structure, how legislation is drafted, the contours through which it flows, the presence of interest groups, political culture and language, and the existence of relevant state agencies and the administrative apparatus—irrespective of the specific policy issue. Accordingly, the specific type of legislation is irrelevant. The names of individual actors, interest groups, and types of agencies may change from one policy arena to another, but the political game remains essentially the same. In addition, some issues, such as school choice, generate much more interest and conflict than others—all the better to expose the hidden dynamics, the backroom deals, unusual coalitions, and strange bedfellows who make up state policymaking.

STATE POLICYMAKING—THEN AND NOW

State education policymaking used to be a good deal simpler than it is today. Throughout much of U.S. history, education policy was shaped by a small handful of key interest groups whose expertise was generally unquestioned. In his comprehensive review of educational politics and policymaking at the state level, Douglas Mitchell observed that, "State education policy making systems were viewed as weak and ineffectual."[5] State lawmakers, overworked and understaffed, did little except appropriate funding for schools. As a result of this limited state capacity, education policymaking was largely the province of education professionals, with conflict limited to disagreements between school board associations, school administrators, and teachers' unions (i.e., within the traditional education establishment).

However, state education policymaking began to change in the 1970s and 1980s as state legislatures were forced by a variety of outside actors and institutions to confront a host of educational challenges, including inequities in school finance, education of students with special needs, and the push for higher standards, assessment, and greater accountability.[6] In the 1980s and 1990s, state involvement increased with federal deregulation, block grants, and various school reform imperatives.[7] Through the "devolution revolution" initiated in the 1980s by Ronald Reagan and continued by his Republican and Democratic successors, power and policymaking have gradually returned to the states.[8] Better schools to support economic growth became the theme of the period, with the states leading the charge.[9] The result has been "an unprecedented growth of state influence over local education."[10] Within the past decade, the education reform movement has been dominated by concerns for higher educational standards, greater accountability for school and student performance, high stakes testing, and increased parental choice.[11]

As a prerequisite for increased state activism, state legislative capacity has increased dramatically in the last four decades, as legislatures have established numerous oversight committees, installed sophisticated data processing equipment, and substantially enlarged their staffs.[12] Passage of the Elementary and Secondary Education Act (ESEA) of 1965 energized state education departments and initiated a much larger role for the state in shaping local education policies. Almost overnight, state education departments were transformed from sleepy little offices into large, active bodies with oversight capability and significant control over local education initiatives. Spurred in part by school finance litigation, as well as a perceived decline in the quality of schooling, state legislatures and, more recently, governors, have become much more active in crafting education policy. Within the past decade, few issues have captured the attention and occupied the time of state legislators more than proposals for school choice. To understand this activism, however, we must first explore the public's interest in choice and trace the birth and growth of the school choice movement in the United States.

Public Support for Choice

Since 1994, public support for school choice has gradually increased within the general population, with support strongest among minorities and the poor.[13] Opposition tends to arise from the traditional education establishment (school boards, administrators, and teachers' unions) and from suburban families whose children already attend good schools.[14] Hubert Morken and Jo Renee Formicola observe that "school choice, as a political movement, is coalescing slowly and unevenly around the United States across racial, religious, economic, moral, and ideological lines," due in part to the diverse composition of school choice advocates, each with different motives and methods.[15] Fully a third of respondents in a Phi Delta Kappa/Gallup Poll indicated that improving public education required finding an alternative to the existing public school system, including various forms of school choice.[16] In the most recent PDK/Gallup poll, nearly one-half of parents (46 percent) favor allowing students and parents to use vouchers to choose a private school at public expense, up from 34 percent just the year before.[17]

Amy Stuart Wells notes that public support for vouchers has increased steadily since the early 1980s.[18] The Texas Poll, a survey of 998 adult Texans' views on education conducted by the University of Texas, found that nearly two-thirds (62 percent) of respondents favored a voucher system for use in public and private schools.[19] A 2002 survey conducted by Zogby International for the Center for Education Reform found that nearly three out of four (72 percent) African Americans supported the idea of allowing poor families to use education vouchers in private, public, or parochial schools; support for the idea declined, but was still high, among Latino and white respondents (64 percent and 62 percent respectively).[20] A poll by Black America's Political Action Committee (BAMPAC) found that 63 percent of African Americans surveyed indicated they would remove their children from public schools and enroll them in charter or private schools if given a choice.[21] In fact, the greatest opposition among the general public to various voucher proposals comes from predominantly white, middle- and upper-class parents who have little to gain (and, perhaps, much to lose) from various targeted choice plans under consideration in the states.[22]

The spread of school choice reflects, in part, deep dissatisfaction with the quality of education in the United States, particularly in urban areas.[23] Despite wave after wave of school reform, the condition of education—particularly in urban areas populated by the nation's most impoverished, disadvantaged children—remains perilous. Reading and math scores in many large cities remain abysmally low. From 1980 to 1990, nearly half of students who entered high school in Chicago dropped out, with little evidence of improvement since then.[24] One-fifth of the public schools in Chicago and New York City are either on probation or in need of immediate intervention.[25] Performance data show wide, persistent gaps in student achievement.[26] Based on data collected by the U.S. Department of Education, the performance gap among minority groups is widening.[27] For example, while the performance gap in reading among 17-year-olds narrowed from 1975 to 1990, it has widened since then.[28]

African Americans and Latinos continue to score below whites on all measures of the National Assessment of Educational Progress (NAEP); this gap is widest in science.[29] The achievement gap between whites and African Americans in math in 4th grade is 29 percent; among 8th grade students, the gap is 29 percent, and among 12th grade students, the gap is 17 percent.[30] Average 12th grade science scores on NAEP reveal a 31 point gap between white and African American students; a 26 point gap exists between white and Latino students.[31] The gap in SAT scores between white and African American students is 201 points; between whites and Latinos, the gap is smaller (135 points) but still significant.[32] The dropout rate among African Americans is nearly twice that of white students; the Latino dropout rate is four times that of white students.[33] On nearly every index of student achievement, "student performance still lags abysmally."[34]

As a result of these persistent gaps in student achievement and the failure of traditional (within-the-system or internal) education reforms to improve student achievement, particularly for the nation's most disadvantaged children, politicians, policymakers, and the general public have been willing to experiment with a wide array of non-traditional, "out of the box" reforms—including state takeovers of chronically failing school systems, mayoral control, the hiring of non-educators (businessmen, lawyers, even former governors) to run school systems, and various forms of school choice (including charter schools, vouchers, home schooling, and contracting out educational services to private companies). Reflecting growing dissatisfaction with the condition of urban education, "public opinion in many cities favors a major overhaul if not outright dismantling of the present educational governance structure."[35]

Clearly, something needs to be done to improve the educational opportunities of at-risk children. Dissatisfaction with repeated failures at educational reform has brought together "an unusual alliance of minority parents, conservative ideologues, and state legislators" searching for ways to reform public education "without totally destroying or abandoning it."[36] One reform that has become popular is the charter schools movement. Increasing numbers of educational reformers view charter schools as an opportunity to provide a more effective education to students who are ill-served by the public school system as it is currently structured—particularly in urban areas.[37] Support for charter schools comes from a disparate collection of groups, including conservatives who also support taxpayer-financed vouchers, business leaders who have lost confidence in nonresponsive, bureaucratic public schools, African American and Latino civic groups, community leaders, and parents who view charter schools as an opportunity to escape failing inner-city schools.[38] In New York City, partners include the Banana Kelly Community Improvement Association, El Puente, the Neighborhood Association for Intercultural Affairs, Allen A. M. E. Neighborhood Preservation and Development Corporation, and the East Harlem Council for Community Improvement. Other choice proposals, such as vouchers, also have their supporters and political coalitions, which will be examined in greater depth in the following chapters.

BIRTH AND GROWTH OF SCHOOL CHOICE

Contrary to popular impression, school choice is neither a new reform, a recent phenomenon, nor an untried idea. In Vermont, home to the longest running school choice program in the United States, the state pays tuition expenses for children attending any public or non-sectarian private school (including schools outside the state) if the local town does not have its own public school. Maine has adopted a similar school choice policy. Most states have enacted various types of school choice plans ranging from interdistrict transfer plans to intradistrict magnet schools.[39] For example, in Jefferson County, Kentucky, the district is divided into clusters of schools, with parents given the opportunity to send their children to any school within a given cluster, with transportation provided by the district.[40] The two largest school districts in Minnesota, Minneapolis and St. Paul, have operated intradistrict public school choice plans since the early 1970s.[41]

Magnet schools, another popular form of public school choice, have been around for decades and were used as a vehicle (albeit an ineffective one) for racial integration.[42] Similarly, alternative schools, designed to reach out to students who do not fit well into traditional schooling, have a long history. New York City operates the largest alternative high school system in the United States. Generally, inter- and intradistrict choice plans generate less controversy because "the number of students who participate is smaller, and public school educators are more comfortable competing with one another than with independent charter schools."[43] The federal No Child Left Behind Act specifies that "children in failing schools should be allowed to attend the school of their choice."[44] In the fall of 2002, students in 8,600 schools nationwide, about 9 percent of public schools in the United States, were eligible to attend the public school of their choice. As Morken and Formicola observe, "Attendance zones for public schools are not the restrictive walls they once were, in law and increasingly in practice."[45]

Recently, officials in the 365,000-student Miami-Dade County school district, the nation's fourth largest, announced that the district will no longer require children to attend their neighborhood schools.[46] Suffering from the defection of more than 10,000 students to charter schools and voucher-accepting schools, the district will be divided into six to eight attendance zones, with parents free to choose any schools within those zones.[47] Currently, 34 applications for new charter schools in the district are pending; if approved, the district could lose another 26,500 students.[48] Superintendent Merrett Stierheim contended that, "I am a fervent believer in public education being the backbone of our way of life, and I'm going to be zealous in trying to protect it. But I'm also pragmatic enough to realize that the issue of choice and competition is here to stay."[49] In Michigan, for example, 15,000 students have utilized an interdistrict choice option.[50] Opposition has arisen mainly from middle-class parents in wealthier, suburban districts, where school officials are concerned about the changing complexion of their districts and the subsequent impact of the migration of poor, often children of color, on their test scores.[51]

Another form of public school choice that has received much publicity in the last few years is charter schools. Currently, 39 states and the District of Columbia have passed charter school legislation; approximately 600,000 schoolchildren are being educated in nearly 2,700 charter schools nationwide.[52] Arizona has the most charter schools in operation, followed by California, Florida, Texas, and Michigan. In Washington, D.C. and Kansas City, Missouri, nearly 14 percent of schoolchildren attend charter schools; about 20 percent of all public schools in Arizona are charters.[53] Nearly 125 charter schools are in operation in Phoenix, Arizona. Charter schools constitute about 2 percent of all public schools nationwide, and represent slightly more than 1 percent of total public school enrollment.[54] Charter schools enjoy wide bipartisan support in Congress. Federal appropriations for charter schools increased from $6 million in 1995 to $145 million in 2000, despite what Joseph Murphy and Catherine Dunn Shiffman characterize as "an often contentious relationship between the Republican Congress and the Democratic White House during those years."[55]

While smaller, voucher programs in various cities have been in operation for a number of years. In the early 1970s, the federal government initiated a voucher program for children from low-income families in Alum Rock, California, although as Robert Bulman and David Kirp observe, the multiple political compromises necessary to create the program transformed it from a voucher plan into an open-enrollment plan of minischools operating within existing public schools (basically, a form of school decentralization).[56] At present, voucher proposals are being developed in nearly half of the country's state legislatures.[57] Privately funded voucher programs, such as CEO America and The Children's Scholarship Fund, operate in approximately 80 cities and serve more than 60,000 students nationwide.[58] Publicly funded voucher programs operate in Milwaukee and Cleveland. Milwaukee's voucher program serves approximately 8,000 students in nearly 100 private schools.

Florida, at the urging of Governor Jeb Bush, operates the only statewide voucher plan currently in existence.[59] The Florida voucher plan is linked to the state's accountability system in which students attending schools receiving an accountability rating of "F" for two out of four consecutive years are eligible to use a voucher at either a public, private, or religious institution. The actual number of students in Florida eligible to use vouchers has been miniscule: for the 2002–2003 academic year, only ten schools in the entire state received consecutive failing grades. Only 45 students used vouchers in the 2001–2002 academic year, although in 2002–2003, that number jumped to 575.[60] In August 2002, a Florida state circuit court judge ruled the state's voucher program unconstitutional; the decision is under appeal. Florida also operates a separate voucher program for children with disabilities. Under the program, special education students can use McKay Scholarships worth anywhere from $4,500 to $21,000 (depending on the student's disability) to pay for private schooling.[61] Approximately 9,000 students have used these scholarships to attend private schools.

Several states offer tuition tax credits and/or deductions for education-related expenses (such as for purchasing computers, tutoring, and in some cases, for private school tuition). Although eligibility and program requirements vary from state to state, Arizona, Illinois,

Iowa, Florida, Minnesota, and Pennsylvania currently offer such programs, while similar legislative proposals are under consideration in several other states. In 1998, nearly 200,000 families in Minnesota received education-related tax relief. A new Florida law "allows businesses to earn tax credits when they donate scholarship money for students to attend private schools."[62] Businesses can "donate up to three-quarters of their state corporate-income taxes to nonprofit groups that give scholarship money for students to attend private schools."[63] Roughly 10,000 students will receive these tax-credit scholarships in the 2002–2003 academic year.[64] Given the highly politicized nature of the debate over school vouchers, policies that manipulate the tax code, offering tax relief for education expenses, are more politically palatable to state lawmakers, although critics have derided policies that allow deductions for private and parochial school tuition as "backdoor voucher" schemes.

Rethinking Theories of Politics and Policymaking

Having traced the birth and growth of the school choice movement in the United States, we now turn to various theories of politics used to explain state policymaking. While there has been an explosion of books about school choice within the past decade, few have attempted in any systematic way to place the politics of school choice within a broader theoretical context. The atheoretical nature of much research on school choice is remarkable given that the movement has dominated the attention of state legislatures, as well as questioned core fundamental values such as freedom, liberty, democracy, equity, and, of course, choice. As Hanne Mawhinney observed, "Despite an explosion of policy research during the past decade that has focused on reform-related issues, the link between theoretical research on policy and the practice of policy making remains weak."[65] While many of the accounts of political battles over school choice are interesting, such accounts do little to advance our understanding of the complex dynamics of state politics and policymaking.

One criticism of traditional political and policy studies in education is that narrow, static, one-shot analyses of policy often lead to inadequate conceptualizations of complex socio-political phenomena.[66] According to Mawhinney, "there is general agreement that current frameworks of policy fail to guide understanding of the politics of policy change."[67] Previous studies that chronicle policy events have been characterized as unsatisfactory "because they lack attention to the basic political structure and processes, thereby providing an incomplete and, frequently, erroneous understanding."[68] Most analyses of policy change are narrow and focus on a single policy or are confined to a single legislative session.[69] However, adequately conceptualizing and understanding the forces that shape policy change can be accomplished only through a longitudinal analysis of policy change across the states.

Effectively understanding these complex forces can be accomplished through more contextual analyses of policy change—in all policy arenas. Any social phenomenon "must be understood in its historical context,"[70] because decision-makers "always draw on past experience, whether conscious of doing so or not."[71] Thus, time may properly be conceived as a stream and events as part of a stream, connected in some way to previous decisions.[72] One of the leading criticisms of traditional policy research is that it is trapped within what I call the *fallacy of presentism*—the tendency in politics and policy studies to ignore the effects of past policies, programmatic, institutional or organizational history, and culture on policymaking. Research studies "tend to focus exclusively on the present rather than past or future trends."[73]

One of the major weaknesses of scholarship in politics and policy studies is that many studies treat their subjects as divorced or somehow apart from their environment. Failure to adequately consider the effects of history and context leads to inadequate, decontextualized conceptualizations of social phenomena—which often conceal the underlying forces, the "grid of social regularities,"[74] that shape and often determine the scope and nature of policy change. Researchers have long been aware of the "structural character and connections of social and educational problems," although they've frequently ignored or dismissed such factors.[75] This book seeks to redress some of the shortcomings of previous analyses by examining the political battles over school choice longitudinally, within the broader organizational and institutional context of the state-level educational policymaking environment.

OVERVIEW AND ORGANIZATION OF THE BOOK

Looking principally at legislative battles over charter schools and vouchers, this book explores the role of interest groups, advocacy coalitions, institutions, political culture, and language in shaping the outcome of school choice reform initiatives. Drawing from insights into organizational learning theory, the book examines the extent to which policymakers utilize research to inform their decisionmaking processes, how (or even if) such considerations affect policy outcomes, and how language is employed to shape the contours of debate over choice. After synthesizing the effects of each of these forces on school choice initiatives, the book concludes by reviewing what has been learned about the political dynamics of school choice and the implications of these dynamics for the future of the movement.

One of the central arguments of this book is that the political dynamics of the school choice movement in states throughout the country tend to be lumped together and treated as similar, when, in fact, significant differences are apparent in the politics of the movement depending on where you look. According to Bryan Hassel, "Despite some common threads, charter school laws across the country differ from one another so greatly that they appear to have been cut from different fabrics altogether."[76] Although the policy rhetoric (arguments for and against charter schools, vouchers, tuition tax credits, etc.) may be somewhat similar, the political dynamics that created strong charter school laws in some states, weak charter laws in others, and pilot voucher plans in Wisconsin, Ohio, and Florida are quite different. This grouping of dissimilar forces is problematic on both a theoretical and a practical, political level because it oversimplifies complex issues—obscuring subtle nuances from state to state.

This book seeks to examine the dynamics of choice in all its multifaceted complexity, highlighting the differential impact of political forces, actors, institutions, and political culture on public policymaking. The analysis is designed to be useful to scholars seeking to develop and validate new theories of state politics, policymaking, and change. By evaluating these theories, the book attempts to strengthen the link between the theory and practice of policymaking. Through better understanding of these forces, policymakers and policy activists will be better able to develop effective strategies of educational reform.

In the chapters that follow, we examine different theoretical explanations and interpretations of often intensely partisan battles to push charter school and voucher legislation through state legislatures. Each chapter of the book examines legislative conflicts over school choice through a different theoretical lens or dimension of the policymaking process. Each chapter begins with an overview of the theory—political culture, institutional theory, interest groups and advocacy coalitions, and organizational learning—followed by examples of the process in action in the United States. From 1997–2000, the author conducted a series of in-depth interviews with key policy elites, including state legislators, committee staff, legislative aides, state education officials, and lobbyists in Texas. The interview data are supplemented with data collected from legislative testimony, committee hearings, floor debates, official documents, fiscal reports, bill analyses, newspaper accounts, and scholarly analyses of the politics of school choice in several other states, with particularly intensive coverage of battles over school choice in Ohio and Pennsylvania—two states that are at the center of heated battles over charter schools and vouchers. Studying intensely partisan battles over controversial policy issues is a boon to scholars and researchers, given the abundance of available sources, commentaries, and analyses.

The examples used in this book highlight the link between the theory and practice of policymaking, demonstrating the utility of each theory as an explanation for policy adoption (or rejection) and change. The use of data from several states highlights the similarities and differences in the politics of school choice and the applicability of our theoretical constructs for explaining that variability across locations. Each chapter concludes with a re-interpretation of the utility of the theory introduced in the beginning of the chapter as an explanation for policy adoption and educational reform.

POLITICAL CULTURE AND LANGUAGE

In chapter 2, we explore the effects of political culture and language on various choice initiatives. Combining Daniel Elazar's classic framework of political culture and recent extensions by Richard Ellis and Dennis Coyle, we examine how a state's political culture serves to constrain (or facilitate) school choice initiatives.[77] Key questions explored in this chapter include: To what extent are policymakers constrained and shaped by the cultural context within which policy is crafted? Are modern-day state policymakers affected by the cultural beliefs of our ancestors? Elazar and others assert that regional differences in political culture are crucial explanatory variables of differential policy outcomes—of the paths and policies chosen by state policymakers. However, in a nation as diverse and highly mobile as the United States, are such explanations of policy outcomes still valid? Or do researchers give too much credence to culture—falsely attributing current policy outcomes to cultural precepts that may not be widely shared by state policymakers today?

INSTITUTIONAL THEORY

Chapter 3 explores the new institutionalism in political science, sociology, and organizational theory and its application to the study of the politics of education.[78] Drawing from the seminal works of James March and Johan Olsen, Theda Skocpol, Stephen Skowronek, and Kathleen Thelen and Sven Steinmo, I use institutional theory as a lens through which to explain differences in the outcomes of school choice initiatives. Key questions explored in this chapter include: How does the institutional structure of state government affect school choice? To what extent is the institutional structure (the "state") merely a neutral arbiter of policy? Or does the institutional structure of state policymaking favor the adoption of some reforms and not others? If so, what does this mean for the formulation, passage, and implementation of school choice initiatives?

INTEREST GROUPS AND ADVOCACY COALITIONS

Chapter 4 looks at the traditional, and still dominant, view of politics and policymaking at the state level—interest group liberalism. Starting with research by David Truman, the chapter examines recent modifications and applications in the neopluralist works of Paul Sabatier, Hank Jenkins-Smith, and others.[79] Key questions in this chapter include: To what degree are different policy outcomes in the states primarily attributable to the actions of interest groups as they pressure policymakers to adopt some education reforms while rejecting others? Do interest groups join forces, creating broad-based (and stable) advocacy coalitions in an attempt to achieve their policy objectives? Or is their influence, like that of a state's political culture, overestimated as an explanation of policy outcomes?

ORGANIZATIONAL LEARNING

In chapter 5, we examine school choice in light of recent developments in organizational learning theory by Peter Hall, Frans Leeuw, Ray Rist, and Richard Sonnichsen,[80] as well as earlier work by Carol Weiss on the use of research to inform policymaking.[81] In business and education, such conceptualizations have become fashionable, as exemplified in the popular writings of Peter Senge.[82] The critical questions explored in this chapter are: Can organizations, such as state legislatures, learn and improve upon past practice? What conditions must exist for such learning to occur? Is the adoption of some school choice initiatives (such as charter schools) and the rejection of others (primarily, vouchers) a product of organizational or institutional learning? Answers to these questions are critically important given the growth and activism of state legislatures in all policy arenas, particularly in education.

Political Dynamics: Lessons Learned

In the concluding chapter, we review what we have learned about the outcomes of the political dynamics of school choice—reassessing the theories of politics and policymaking explored in the previous chapters. Included in this chapter is a discussion of political learning—how, over time and through experience—various interest groups have learned how best to manipulate the political game to achieve their objectives. We explore the mistakes and missteps, as well as the clever strategies and successes, of advocates and opponents of school choice and analyze ongoing conflicts over choice in the American states. The book concludes with an assessment of the prospects of expansion of school choice initiatives in the United States, particularly in light of the recent U.S. Supreme Court ruling upholding the constitutionality of the voucher program in Cleveland.

A Final Word

Before beginning our journey into the fascinating world of state politics and policymaking, a gentle reminder is in order. This book utilizes a multidisciplinary approach, drawing on an extraordinarily diverse literature in the fields of history, organizational studies, political science, sociology, and education to help readers better understand the politics of education policymaking in the American states. Such conceptual borrowing is not new—in fact, it is quite common in an applied field such as education. Scholars and writers in the field of education routinely look to advances in other fields to inform their understanding of educational processes and policies. However, a disturbing tendency among scholars in more established disciplines is to ignore the seminal contributions made by those within education to the study of educational politics and policymaking. Such one-way "borrowing" or unidirectional inquiry is unfortunate and only serves to limit our understanding of state politics and policymaking. This book seeks to combine the best scholarship in the politics of education with advances in other fields, crafting it into a more coherent, integrated whole. The result, we hope, is stronger theoretical coherence across disciplines and a more comprehensive, contextualized understanding of the politics of school choice in the American states.

Chapter 2

Cultural Dynamics: Political Culture and Language in Policymaking

At its core, policymaking is a relational process situated within a specific, nested context, involving a series of complex interpersonal interactions—some collegial, some confrontational. The nested context within which policymaking takes place—the intensely political environment of a state capitol—sets the parameters of policymaking, shaping the contours within which this relational process operates, with important implications for policymaking. Forty years ago, in his seminal, if oft-misused and little understood, *The Semi-Sovereign People*, E. E. Schattschneider issued a radical statement about the power of agenda-setting when he asserted, "the definition of the alternatives is the supreme instrument of power."[1] Schattschneider did not argue that the *selection* of a preferred option was the ultimate test of power, but rather that the *process* by which certain policy options are considered within the range of acceptable alternatives, while others are not, is the ultimate test of power. Since policymaking is fundamentally an interpersonal, relational process, we begin by examining the cultural context within which policymaking takes place.

Political Culture

Beginning with the pioneering work of Daniel Elazar, political culture is defined as the "political thoughts, attitudes, assumptions, and values of individuals and groups."[2] It is a type of "mind set" that limits the range of policy alternatives considered.[3] In any culture, most people will "take for granted a particular course of action or consider only a few alternatives."[4] As a result, culture plays a significant role in shaping how problems are defined and solutions adopted.[5] Christopher Bosso argues that "the 'received culture' seems to have had a greater role in defining the range of legitimate alternatives than any policy elite or interest group."[6] Political culture helps to define "the range of acceptable possible alternatives from which groups or individuals may, other circumstances permitting, choose a course of action."[7] While political culture rarely determines behavior on specific issues, "its influence lies in its power to set reasonably fixed limits on political behavior and to provide subliminal direction for political action in particular political systems."[8] The key research question is whether one political culture exists to which most Americans adhere, or whether different sociocultural values and, by extension, different political cultures, are evident from state to state or region to region. Would different sociocultural values affect state policymaking and, if so, how?

According to Elazar, the national political culture of the United States consists of a synthesis of three distinct political subcultures: individualistic, moralistic, and traditionalistic. Each political subculture is the product of distinct historical immigration patterns. Scholars believe it is a historical fiction to assume that each wave of immigrants who came to the shores of the United States shared common cultural values. On the contrary, differences in political cultures arise out of still-pertinent sociocultural differences among immigrants. Drawing from these immigration patterns, Elazar asserts that each state is distinctive "in terms of the composition of its people, its political goals and processes, and the way in which its people respond to those goals and processes."[9]

Individualistic Political Culture

In the individualistic political culture, politics is treated as a market in which public policy serves strictly utilitarian purposes. Private interests dominate policymaking, with limited government intrusion into the private sphere. Treating politics as a business, Elazar asserts that "public officials, committed to 'giving the public what it wants,' are normally not willing to initiate new programs or open up new areas of government activity on their own initiative. They will do so when they perceive an overwhelming public demand for them to act, but only then."[10] Reflecting the sociocultural values of English, continental, Eastern European, Mediterranean, and Irish immigrants, this political culture is dominant in the mid-Atlantic states of New York, Pennsylvania, Ohio, New Jersey, Indiana, and Illinois, among others.

Moralistic Political Culture

Citizens inculturated into the moralistic political culture believe public policy should serve the common good, for the purpose of creating the good society. Ultimately, political conflict "has been and continues to be animated by fundamentally different visions of the good life."[11] Rooted in the cultural values of the Puritans, Jews, Scots, Dutch, Scandanavians, and Swiss peoples, the moralistic political culture encourages a proactive approach to policy initiation. According to Elazar, public officials operating within a moralistic political culture "will seek to initiate new government activities in an effort to come to grips with problems."[12] Unlike the individualistic political culture, politics is viewed as an affirmative process, leading to the betterment of society for all.

Traditionalistic Political Culture

The traditionalistic political culture, with its roots in Southern agrarianism, adopts an elitist, paternalistic attitude toward politics and participation in the political system. Special importance is placed on social and familial ties—this culture is evident among Latinos as well as Southern whites. Ambivalent toward the market, the purpose of politics is to preserve the social order. Politics and, by extension, public policy, is conservative and custodial, rather than proactive or initiatory. Engrained with a virulent anti-bureaucratic philosophy, government action is constrained to a degree unparalleled in the other political cultures. Since open conflict is suppressed within this political culture, interparty competition is discouraged; conflict is limited to intraparty factionalism. Note, for example, the dominance of the Democratic Party in Southern politics from Reconstruction until 1980. For nearly a century, the South was controlled by the Democratic Party; only in the past two decades has the Republican Party become competitive in Southern politics.

Adding to the complexity of three distinct political subcultures existing within the United States, Elazar observes that intergenerational migration patterns have produced situations in which several states are a combination of (usually two) political cultures, with one dominant and the other subordinate. For example, 16 states are classified by Elazar as either traditionalistic, traditionalistic/moralistic, or traditionalistic/individualistic; 17 states are classified as either individualistic, individualistic/traditionalistic, or individualistic/moralistic; another 17 states are either moralistic or moralistic/individualistic. No state has a dominant moralistic political culture with strong traditionalistic undertones. The existence of powerful, subordinate subcultures competing with the dominant political culture in a state produces interesting political conflicts over the nature, scope, and direction of policy initiatives.

POLITICAL CULTURE AND STATE EDUCATION POLICY

Cultural and historical forces help shape the context within which education policymaking occurs. Thomas James observes that "varying political cultures are significant in structuring attention to educational issues at the state level."[13] Together with institutions, culture shapes the nature and direction of policy change. For example, states that have a reformist or progressive tradition in their political culture tend to be leaders in policy innovation.[14] Political culture becomes incorporated through formal institutions and informal processes into the policy system.[15] It is through institutions, together with the values, attitudes, and beliefs of participants, that political culture exerts its effect on policy change.

Surprisingly, few studies applying Elazar's political culture framework have been conducted in education. Educational researchers have generally been much more interested and concerned with studying organizational culture at the school or district level, rather than political culture at the state or national level. Catherine Marshall, Douglas Mitchell, and Frederick Wirt use differences in state political culture to explain variance in policy outcomes.[16] Drawing from policy documents, observations, and interviews with policy elites in Wisconsin, Illinois, California, Arizona, West Virginia, and Pennsylvania, the researchers examined the cultural influences shaping state education policy. Testing Elazar's cultural framework, the researchers found that the framework explained some of the variation among states, but its explanatory power was limited by a series of national and international forces sweeping through the states, which produced similar state policy responses.

Donal Sacken and Marcello Medina used the political culture taxonomy developed by Mitchell, Marshall, and Wirt as a framework to analyze bilingual education legislation in Arizona.[17] The researchers found that Arizona's traditionalistic political culture served as a significant constraint in the formulation and passage of bilingual education legislation, producing legislation far less expansive and comprehensive than originally conceived. Maenette Benham and Ronald Heck conducted a historical study of the relationship between state political culture and policymaking in Hawaii over a period of 150 years.[18] The researchers found that the unique characteristics of Hawaii's political culture, which Elazar characterized as traditionalistic, reflect limited citizen involvement with an overriding concern for the value of efficiency. Their findings provide support for Elazar's cultural framework. Generally, studies of policy restraint situated within the traditionalistic political culture tend to be supportive of Elazar's framework, since that political culture is unambiguous in its embrace of limited government intervention and intolerance of expansive public policies.

A number of studies of the effects of political culture on higher education policy have also been conducted. Peter Garland and S. V. Martorana analyzed the impact of political culture on state legislation affecting community colleges. The authors found that differences in state political culture are associated with variation in leadership, participation, and involvement.[19] Elaine Freeman conducted a comparative analysis of the effects of political culture on higher education policy in Missouri and Oklahoma.[20] Using Elazar's typology, Missouri has an individualistic political culture, with strong traditionalistic undertones; Oklahoma has the reverse—a traditionalistic political culture, with strong individualistic undertones. Freeman concluded that each state's higher education policies reflect differences in the political cultures of the states—with Missouri's higher education policies demonstrating a concern for marketplace issues, while Oklahoma's higher education policies were dominated by policymakers' interest in preserving the status quo.

Applying Elazar's political culture framework to the micropolitical (local) level, Lorn Foster attributed variation in the beliefs of school board members in Nebraska and Louisiana to differences in each state's political culture.[21] In another study, Belinda Pustka analyzed the effect of political culture on school district policymaking in Texas and found that political culture influences policy decisions concerning students' placement in special education, vocational programs, and district spending.[22] A series of related studies attempt to situate political culture within the nexus of religion and educational policymaking at the county and school district levels, exploring such questions as: "How well do reform initiatives such as outcomes-based education, site-based management, and the movement toward integrated social services in schools mesh with the dominant political culture of the district, region, or state? As religion relates to the larger political culture, to what extent may differences among districts . . . be attributable to the effect of religion on the political culture?"[23] While results of studies testing the operational efficacy of political culture at the micropolitical level are mixed, few dispute the assertion that more general cultural values permeate the policymaking process at the state level. The question becomes how these cultural values manifest themselves in debates over controversial issues such as school choice and how value differences among policy elites shape the contours of those debates.

POLITICAL CULTURE AND SCHOOL CHOICE

Can Elazar's theory of political culture explain variation in state policymakers' responses to school choice? Charter school legislation in most states gives preference to at-risk students; many state charter school laws explicitly encourage activists to create schools to better serve the needs of such students. With respect to charter schools, then, the policy rhetoric suggests that concerns for equity are a driving force of the movement.[24] Applying Elazar's cultural framework, states with a moralistic political culture, with its proactive emphasis on creating a more equitable society, should be leaders in the charter school reform movement.[25] In fact, six of the first eight states to pass charter school legislation have moralistic political cultures.

For example, Minnesota, the first state to pass charter school legislation, has a moralistic political culture, suggesting that state policymakers would be proactive, constantly initiating new programs and reforms. Indeed, Minnesota has long been in the vanguard of the educational reform movement and boasts a long tradition of progressive education reform. In 1958, the state legislature "piloted a plan that allowed at-risk students to exit their own neighborhood districts to obtain remedial instruction in other public and private schools."[26] That same year, lawmakers adopted a proposal allowing "high school juniors and seniors to take courses at public and private colleges," a policy now in use throughout the country.[27] In 1991, the legislature passed the nation's first charter school law, which appears to give validity to Elazar's theory. Wisconsin, which also has a moralistic political culture, is home to charter schools and to the nation's first state-approved publicly funded voucher program in Milwaukee.

However, the difficulty with using political culture to explain state policymakers' preference for charter schools lies in the fact that most states (39) and the District of Columbia have passed charter school legislation. Many of these states have traditionalistic or individualistic political cultures. Texas is a very conservative state with a traditionalistic political culture; yet in 1995 the state legislature passed what was, at the time, one of the strongest charter school laws in the nation, a fact seemingly inconsistent with Elazar's theory, wherein state legislators play a "conservative and custodial" role.[28] Pennsylvania, which passed a stronger charter school law than either Texas or Ohio, has long been guided by an individualistic political culture. The Pennsylvania state legislature is dominated by professional politicians whose decisionmaking style is best characterized as one of "bargaining, compromise, and log-rolling rather than by ideological appeals [or] good government crusades."[29] In his study of state education politics in Pennsylvania, Ellis Katz quoted one source who stated, "If you're not willing to broker, then you're going to be unsuccessful. Brokering is at the heart of Pennsylvania politics. Especially with the partisan division, you have to come to compromises."[30] This penchant for compromise "is the legislature's solution to the deep partisan, geographical and ideological cleavages which could paralyze the legislature were they allowed to surface on every issue."[31]

Arizona's charter school law, one of the strongest in the nation, reflects the "entrepreneurial and anti-bureaucratic philosophy of the wide open West."[32] Arizona has long been considered "one of the most politically conservative states in the nation."[33] Under Elazar's typology of political culture, Arizona is a traditionalistic state, with a strong current of moralism.[34] In a similar vein, Ohio's traditionally conservative, individualistic political culture, reflected in a generally conservative press, leads to annual rituals of "no new taxes" pledges by state politicians—regardless of the condition of the public schools.[35] In her longitudinal study of education funding in Ohio, Linda Bennett found that "state culture, its continuing existence and perpetuation by professional politicians, is at the heart of Ohio's education policy process."[36] Since Ohio is classified as having an individualistic political culture, we would not expect much policy activism in the area of school choice. However, not only does Ohio have charter schools, but state lawmakers passed voucher legislation as well.

One could argue that states in which moralism is strong but not dominant in the political culture would be more likely to adopt charter school legislation than states absent such tendencies. The political cultures of Ohio and Arizona contain strong strains of the moralistic political culture. However, the existence of a moralistic political culture, however strong, is an inadequate predictor of a state's likelihood to adopt school choice legislation. Bryan Hassel analyzed the effect of state political culture on charter school legislation and found that states with moralistic political cultures "were more likely than other states to pass charter laws (and to pass strong ones) by January 1996, but the differences were not statistically significant."[37] Furthermore, of the eleven states without charter school legislation, seven (64 percent) have dominant moralistic political cultures. Thus, Elazar's typology of state political cultures does not provide much guidance in predicting which states are most likely to be leaders in the school choice movement. In fact, the most that could be said about the typology is that states with moralistic political cultures are more likely to be policy leaders in the education reform movement *writ large*—whether the reforms are liberal or conservative.

Is it possible that school choice reflects the individualistic rather than the moralistic political culture? A plausible argument could be made that school choice, in its embrace of individual freedom, choice, market efficiency, and accountability, reflects individualistic attitudes toward the purposes and functions of education. However, even more difficulties arise with using the individualistic political culture as a variable to explain state adoption of school choice reforms. Of the first eleven state legislatures that passed charter school legislation, none had an individualistic political culture. The Wisconsin legislature, responsible for passing Milwaukee's voucher plan, operates within a moralistic political culture. Florida, the first state to pass a statewide voucher plan, is dominated by a traditionalistic political culture. Only Ohio, home of Cleveland's voucher plan, has an individualistic political culture. Hassel concluded that "all that one could safely predict is that states with traditionalistic cultures would shy away from charter laws altogether, and from strong charter laws in particular, in light of their disinclination to experiment with system-changing reforms," although this interpretation fails to explain why three of the top five states with the most charter schools (Arizona, Florida, and Texas) have traditionalistic political cultures, with scarcely any degree of the moralistic strain in either Florida or Texas politics.[38]

Although differences in state political culture may not be reliable predictors of legislative activism in school choice, few would deny that the United States has a distinctive, generalized political culture that most Americans adhere to and support. While this political culture embraces sometimes conflicting values, such as individualism and community or equity and choice,[39] certain cultural markers are clearly evident in the language and rhetoric of public policies.

Since policymakers are influenced by their own political values and beliefs, public policies reflect the political culture within which policy is made. In the next section, we explore how political culture and language intersect, and the way in which this relational process shapes education policy.

POLITICAL CULTURE AND THE POLITICS OF LANGUAGE

Drawing on work by E. E. Schattschneider, Ludwig Wittgenstein, and Murray Edelman, the struggle to define the range of alternatives, a form of agenda setting,[40] may be viewed as a type of language game in which rhetoric is used to frame the parameters of policy debate, leading to the privileging of some options in the policy discourse over others.[41] Politicians, lobbyists, and policymakers in all arenas regularly play language games, capitalizing on the definitional ambiguity inherent in the policy process, particularly in jargon-laden fields such as education. For example, what do policymakers mean when they engage in debates over centralization versus decentralization or debates over school choice?[42]

The language games played by policymakers in this definitional struggle play a key role in limiting the range of alternatives under consideration at any given time as well as privileging certain options within policy discourse. These language games are culturally conditioned by the political context within which the games occur. Analysis of the rhetoric used in debates over school choice is particularly illustrative in highlighting the facets of language games since no recent state-level educational reforms generate as much conflict or hortatory political rhetoric as proposals for choice.[43] By analyzing the rhetoric used by policymakers in debates over school choice, I demonstrate how political culture and language are used to frame debates over choice, defining and limiting the range of acceptable alternatives, and influencing legislative outcomes.

Language games are predicated upon the notion that language is not and can never be politically neutral or incontestable. Edelman observes that "language is the key creator of the social worlds people experience, not a tool for describing an objective reality."[44] Truth is simply a property of linguistic entities, of structured sentences.[45] As Richard Prawat and Penelope Peterson assert, "Truth, in the strictest sense, is a successful move in the 'language game.'"[46] Given the nonobjectivity of language, "virtually every word and phrase used in casual speech and thought bears a heavy connotative burden which opens the way to socially approved conclusions and inhibits the recognition of possibilities that are not culturally condoned."[47] Language, particularly metaphors and analogies, is used "to control people's evaluations of policy alternatives."[48] It is the medium through which policy options are interpreted and evaluated, and preferred alternatives advanced.[49] Thus, how an issue is defined or redefined influences "the probability of a policy outcome favorable to advocates of the issue."[50]

Language structures decisionmaking insofar as it can favor a particular result and diminish the consideration or adoption of alternatives.[51] How is this favored result achieved? Language is employed either to expand or narrow the scope of conflict since "every change in the scope of conflict has a bias; it is partisan in its nature."[52] This bias manifests itself in the "exploitation of some kinds of conflict and the suppression of others."[53] Edelman argues that the critical element in political maneuvering for advantage is:

> the creation of meaning: the construction of beliefs about events, policies, leaders, problems, and crises that rationalize or challenge existing inequalities. The strategic need is to immobilize opposition and mobilize support. While coercion and intimidation help to check resistance in all political systems, the key tactic must always be the evocation of interpretations that legitimize favored courses of action and threaten or reassure people so as to encourage them to be supportive or to remain quiescent.[54]

Linguistic representations construct reality and "political language is a major factor in the formation of political thinking."[55] According to Elazar, "language is the 'program' through which humans register and structure external reality; hence language is a major element in the creation of different sensory worlds and the formation of thought."[56] Language conveys intentionality through its power to create desired outcomes and to influence relationships through shared social meanings.[57]

Rhetoric focuses on the use of language—how it influences the thoughts and conduct of an audience.[58] Some words, like equality, possess rhetorical power, signifying potent human ideals.[59] Sources of rhetorical force reflect cultural, psychological, and historical influences.[60] The choice of words used, and the care with which they are crafted into policy arguments, open windows of opportunity for some actors or groups, while closing off others from policy discourse. Schattschneider observed that "a conclusive way of checking the rise of conflict is simply to provide no arena for it."[61] In policy debates, the structuration of language confers legitimacy since, "It is language that confers status on reality, not reality on language."[62]

Meaning is derived from its use in language games, which is to say that meaning is dependent upon the way in which issues are framed in political discourse.[63] In his study of presidential campaigns, William Riker found that "campaigns are rhetorical exercises: attempts to persuade voters to view issues in the way the candidate wishes them to."[64] After all, "Rhetoric is the art of persuasion."[65] And as Noreen Garman and Patricia Holland assert in their analysis of the rhetoric of school reform reports, "rhetoric matters."[66] It shapes practice insofar as it influences the thoughts and actions of the audience. Arguments appeal to emotion, rather than reason, and the sacred remains "beyond challenge or question."[67] Once ideas become axiomatic in policy discourse, refutation becomes nearly impossible. As Ann Norton argues in her exploration of symbolism in political culture, "The force of a political idea lies in its capacity to transcend thought" becoming "a set of principles unconsciously adhered to, a set of conventions so deeply held that they appear (when they appear at all) to be no more than common sense."[68]

Problematizing political discourse and the context within which it takes place enables researchers to better understand how language games shape educational policymaking. For example, David Tyack and Elizabeth Hansot documented the success political progressives of the early twentieth century had in removing public control from educational decisionmaking and placing the process in the hands of "experts."[69] By defining spheres of activity and policy as beyond the political realm, or as nonpartisan, professionals were able to insulate key areas of educational decisionmaking from popular pressures.[70] Similarly, by contracting the sphere in which lay people could participate, school professionals were able to implement their policy solutions largely without opposition, made possible through the political manipulation of language.[71] By redefining partisanship as a form of deviancy, partisanship was sanitized out of political discourse, leaving nonpartisanship or expert control as the preferred solution. By manipulating language in this manner, school professionals were able to exclude the voices of the public from the policy dialogue. Thus, far from being a neutral transmitter of ideas, language serves to structure policy debates, privileging some policy alternatives while excluding others. To illustrate this process, we now examine language games played by politicians, lobbyists, and activists in battles over school choice.

PLAYING LANGUAGE GAMES WITH SCHOOL CHOICE: A CASE STUDY OF POLITICAL RHETORIC

In Texas, the debate over school choice centers around three distinct yet interrelated language games—the nature of the state, the market, and choice (defined within a framework of individualism, liberty, and equal educational opportunity). These categories were derived from the rhetoric employed most frequently by policymakers in debates over charter schools and vouchers. The outcome of these language games influenced the state legislature's decision to adopt charter school legislation while repeatedly rejecting voucher proposals. A close examination of the battle to control the rhetorical scope of the conflict within these games highlights this process.

Language Game #1: The Nature of the State

The first language game centered on the nature of the modern state, particularly its expansive, bureaucratic tendencies. Critics lambasted the bureaucratic inefficiency of public schools and of government as a public entity, generating calls for reform. The debate was framed within the context of freeing schools from encroachment by the state. In an editorial in support of charter schools, Joe Christie, former state senator and co-founder of the Children's Education Opportunity Foundation, argued that public schools need to be "free from the 13 pounds of rules and regulations that restrict a teacher's ability to respond to each child's individual needs. Free from the bloated school bureaucracy that, according to the Texas Research League, siphons off 42 cents of every dollar spent on public education in Texas."[72] Republican state Representative Kent Grusendorf, a member of the House Committee on Public Education, criticized the current system as "overburdened with rules and regulations."[73] School choice, he asserted, "will set good teachers free to teach, and curb the worst excesses of school bureaucrats. More educational dollars will find their way to the classroom."[74] Former Governor George Bush agreed, calling for relief from overburdensome laws, rules, and regulations in the educational system.[75]

This view was held not only by Republicans and other conservative policymakers but also by many liberal Democrats as well. Then-Governor Ann Richards, a Democrat engaged in a tough reelection campaign against Republican challenger George Bush in 1994, also adopted anti-statist rhetoric, asserting, "We have to free up our schools and create an atmosphere where the local schools feel free to do whatever is necessary to help their students succeed."[76] During debate on the House floor over a charter school proposal, a representative stated, "This concept [charter schools] is consistent with all the trends going on in our society with respect to deregulation and freeing people to do what they know needs to be done." This statement mirrored earlier debate in both the House and Senate education committees. As one representative stated amidst debate in the House Public Education Committee, "I want to move forward and give control of my schools back to my parents." Another remarked, "If you deregulate, if you free people to do what they know is right rather than try to control their actions from some centralized power structure, they will not let you down."

The rhetoric employed in the first language game reveals little disagreement about the bureaucratic inefficiency of public schools and, by implication, the presumed low performance of public schools. Criticism of the public school system has long been common among Republican legislators. However, they were joined in their criticism by both liberal and conservative Democrats in the state House and Senate. The "dialogue" was decidedly one-sided, so one-sided in fact that both Republicans and Democrats seemed to agree on the need for reform. This consensus is particularly striking given evidence that performance on state exams in Texas had been steadily improving.[77] By framing the language game and thus the public debate around the issue of freeing the public schools from bureaucratic encroachment by the state, the question became not whether public education in Texas *should* be reformed, but rather *how*. The scope of the conflict within this language game was very narrow, as both Republicans and Democrats in the state legislature and executive branch agreed that the public school system as a statist institution was bureaucratic, inefficient, and in need of reform.

Interestingly, the language game focusing on the nature of the modern state did not privilege charter schools over voucher plans, or vice versa. The language game provided no arena for asserting the superiority of either reform within the nested context of this particular language game. Rather, it merely favored action *qua* action, effectively privileging both charter school and voucher reforms. The outcome of this first language game was a political preference for action—the pressing need to reform bureaucratic-laden public schools by restricting the grasp of the state and returning to local, parental control.

Language Game #2: The Market

The second language game focused on the nature of the market. Unlike the rhetoric of the first language game, the second questioned the fundamental assumptions of the public and private spheres—leading to an expansion of the scope of the conflict and a corresponding battle over the merits of charter school and voucher plans. Proponents of school choice argued that choice would improve education through the principles of the free-market.[78] Representative Ron Wilson (D), representing an urban district where "nearly 87 percent of high school seniors failed a statewide achievement exam," introduced a bill during the 1993 legislative session that would have created a "voucher program for disadvantaged students. Under the pilot, parents could use public tax dollars to pay tuition at a private school."[79] Wilson supported charter schools as well, arguing that they introduce "badly needed competition and innovation for public schools."[80] His position mirrored that of Governor Bush who stated, "competition is something we shouldn't be afraid of... charter schools inject competition into the public education system."[81] Several newspaper editorials touted the virtues of charter schools as a way to "provide parents with options within the public school system and create a form of competition that can elevate academic performance."[82]

Somewhat surprisingly, little dispute arose over the "fact" that charter schools would encourage competition within the educational marketplace. Accordingly, little organized opposition accompanied the creation of charter schools in Texas. Governors Richards and Bush, Lieutenant Governor Bullock, other top legislative leaders, and numerous interest groups, including the Texas Parents and Teachers Association (PTA) and the Texas State Teachers Association (TSTA) supported various charter school proposals. Testimony by numerous witnesses in hearings before the House Public Education Committee and the Senate Education Committee revealed little disagreement on the creation of charter schools. In Texas, the debate centered not upon whether charter schools should be created, but rather how many should be authorized by the legislature.

However, similar rhetorical arguments about the benefits of voucher plans operating within the market system met stiff resistance. During an interview, a senior official with the Texas Elementary Principals and Supervisors Association (TEPSA) stated, "It seems to be very much a right-wing agenda. You also have folks that are very much pro-business, kind of this free-market competition approach that will improve our schools." Senator Gonzalo Barrientos (D), a voucher opponent, remained unconvinced of the virtues of applying free-market principles to public schools. He stated, "I'm sure there would be good people and good schools. But there also would be some fly-by-nighters who will take the money and run."[83] Voucher critic Brad Duggan, executive director of TEPSA, stated, "It's real clear what happens when you give someone a check and tell them go out in the marketplace and shop for services. Parents and students get taken advantage of."[84] An editorial argued that "not every child can take advantage of vouchers because there are only so many schools and so many desks."[85]

Paul Sadler (D), chair of the House Public Education Committee during the 1995 legislative session, stated that "we seem to be sending a message to the public . . . that we are going to let them take their money and run. I'm not willing to send that message."[86] Carolyn Boyle, spokesperson for the Coalition for Public Schools, which helped defeat a voucher bill in 1995, argued that "Texas can't afford a $1 billion private school voucher experiment that would drain money from underfunded public schools."[87] Senator Barrientos concurred, stating "vouchers will not improve teaching methods; vouchers will not repair outdated buildings; vouchers will not make parents more attentive or involved in their children's schoolwork. Vouchers will undermine all these things by taking money out of the system."[88]

Unlike the first language game in which the scope of the conflict was narrow, the language game of the market privileged the adoption of charter school proposals over voucher plans. Charter school proposals were viewed as "safe competition" insofar as they were framed within the nested context of the public education system. Because charter schools are public schools, the benefits of competition via the marketplace within the public sphere was not challenged since the net effect (transfer of funds from one public entity to another) was neutral. Since the language game of the market with respect to charter schools did not threaten the viability of the public sphere, conflict was minimized and charter proposals received strong bipartisan support. This support disappeared when the market metaphor of this language game was applied to voucher proposals.

Language Game #3: Choice

Debate in the third language game centered on choice, which evoked the values of individualism, liberty, and equal educational opportunity. Interestingly, equal educational opportunity was invoked both by supporters and opponents of school choice. Within the language game of choice, the scope of the conflict most expanded, particularly with respect to vouchers, while much of the charter school debate supporting freedom of choice followed well-established and predictable patterns. Governor Richards, herself a former schoolteacher, stated "parents should have the right to select the public school that is best for their child. If the right school doesn't exist, they should be able to create one."[89] Many African American parents in Texas pushed for open-enrollment charter schools,[90] according to Reverend Frank Garrett, Jr., chair of the Coalition for Quality Education—a group that supports quality education initiatives for African Americans.[91] Similar arguments in favor of choice were made by supporters of voucher plans. Speaking in support of vouchers, Representative Wilson stated "we're talking about the people who fund the system having the option to take some of their tax dollars and apply them to their child's education at a private institution."[92]

In Texas, choice vis-à-vis vouchers was framed within the often-conflicting rhetoric of individualism, liberty, and equal educational opportunity. Voucher proponents often combined all three in an attempt to control the policy discourse. Legislators advocated vouchers as a method to "equalize educational opportunity by giving all families the options now enjoyed only by those wealthy enough to send their children to private schools or move to areas with high-quality public schools."[93] Wilson passionately argued, "I live in the middle of the black community in Houston . . . I see those kids having to negotiate death and destruction and drugs every day to get to school. They've got no way out."[94] Jimmy Mansour, representing an organization called Putting Children First,[95] argued that "over 500,000 Texas schoolchildren today are required to attend failing schools—that is just wrong."[96]

In an editorial, Representative Kent Grusendorf (R), a member of the House Public Education Committee who introduced a voucher plan for low-income students during the 1993 legislative session, argued, "It makes absolutely no sense to mandate that a child should attend a certain school just because some elitist drew a line on the ground and said if you live on this side of the line, this is your only choice."[97] Senator Jane Nelson (R), like Wilson and Grusendorf a supporter of a pilot voucher program for disadvantaged students, argued that "the bottom line is the (voucher) recipients would be children of low-income families . . . in inner-city school districts, who have no other way out of the system."[98] An African American parent stated, "What we're saying is that our tax dollars can go toward private education, and it should be the parent's choice."[99] Representative Paul Sadler observed that "the intent of the legislation was to give parents as much flexibility as possible in where their children will be educated, particularly in school districts with a history of low performance."[100]

However, other policy elites disputed the claim that vouchers would provide equal educational opportunities for disenfranchised children. In an interview, an official with the Texas Elementary Principals and Supervisors Association said, "If you look at the Christian Right, things that are put out by certain folks, their mission is to eliminate the public schools. It really is." She observed, "You have motives. There are people involved that simply want individual rights. That puts public education in a different... it gives it a different purpose than what its had." In testimony before the Senate Education Committee, Linda Bridges of the Texas Federation of Teachers commented on the attractiveness of school choice, stating, "We understand the political appeal of choice as a concept. It is always smarter to campaign for choice than for no choice." Members of the Texas State Teachers Association expressed concern that choice programs could further widen the gap between rich and poor.[101] TSTA President Richard Kouri sharply criticized voucher plans, calling them the "agenda of the extreme right and other ultra-conservative groups," to which Representative Wilson retorted, "I can hardly be classified as part of the extreme right. Educators still have their heads stuck in the sand."[102]

Voucher plans were criticized by several minority lawmakers as returning segregation to public schools—making them separate and unequal. Magnolia McCullough of the African Methodist Episcopal Church for the 10th Episcopal District expressed the fear that initiatives for school choice could lead to further segregation of schools. She stated that "it could be viewed as re-segregation. Good teachers are going to follow the money and the best students."[103] Brad Duggan, executive director of TEPSA, criticized the motives of voucher proponents, claiming that "the whole voucher program has been sold as [the] way for low-income parents to obtain the same quality of education as the governor's children. But we know private schools aren't going to open their doors to all students."[104] A. P. Brooks, education reporter for a major newspaper in the state, observed that "the discourse in the Legislature [on vouchers] is divided largely along racial and ethnic lines."[105]

Most minority lawmakers in the state legislature opposed voucher plans amid fears of widening the division between the haves and have-nots.[106] Expressing concern over vouchers, Senator Gregory Luna (D) said, "I hope we're not extracting from the system the problemless [students] and leaving behind the others... who need our help the most."[107] Representative Sylvester Turner agreed and argued that, "In the foreseeable future, the overwhelming majority of

African American and Hispanic children will be educated in the public school system. Every dollar sent to the voucher system and private schools is a dollar less from the public school system. If we work to destroy that base, we will be condemning those children to an inferior education."[108] Gil Gamez, director of the League of United Latin American Citizens, said, "Hispanics are waking up to the fact that so-called reform proposals are only disguised efforts to shut out minorities from educational and economic advancement."[109] Senator Royce West (D), an African American, predicted that "this [voucher bill] will create flight from the inner city."[110]

In the language game of choice, the rhetorical scope of the conflict was most expanded, encompassing a wide array of arguments on behalf of choice. Choice proponents invoked the cultural referents of freedom, individualism, liberty, and equal educational opportunity, cleverly linking these referents to the language game of the market. The widened scope of the conflict within this game privileged charter school proposals over voucher plans. Like the language game of the market, charter schools proposals were viewed as "safe choice," successfully packaged as offering the benefits of choice while ensuring equal educational opportunity. While not denying the appeal of educational choice, opponents of voucher plans played upon the policy elite's fear of unregulated, unrestrained market forces and framed voucher plans as creating more educational inequity for children, effectively countering the equal educational opportunity arguments made by voucher proponents.[111] By expanding the scope of the conflict within the policy discourse, consensus on vouchers was unattainable. As a result, the language game of choice facilitated the adoption of charter school legislation while marginalizing voucher plans.

Language Games, Political Culture, and the Structure of Political Discourse

Earlier in this chapter, I noted Edelman's argument that the critical element in political maneuvering for advantage is "the creation of meaning: the construction of beliefs about events, policies, leaders, problems, and crises that rationalize or challenge existing inequalities . . . the key tactic must always be the evocation of interpretations that legitimize favored courses of action."[112] The battle over charter schools and vouchers may be viewed as a series of language games over the creation of meaning within the context of state legislative policymaking. To achieve success, language had to be manipulated to limit or expand the scope of conflict, thereby influencing the adoption of preferred policy options. Analysis of the language games played by legislators, staff, lobbyists, and others reveals the existence of three interrelated language games within the policy discourse: (1) the nature of the state; (2) the market; and (3) choice. In each of these language games, debate centered on the evocation of interpretations that privileged one course of action, one policy option, over others. By using language, particularly metaphors and analogies, to structure the debate over school choice, policy elites were able to encourage the Texas legislature to adopt charter school legislation while blocking voucher proposals. This outcome is consistent with research by Deborah Stone and William Riker that demonstrates how language is used to control the evaluation of policy options.[113] The key to success was the linkage of political rhetoric in the policy discourse to dominant cultural values in the political culture.

Policy change is difficult when it conflicts with strongly held public values in the political culture.[114] For example, Ann Bastian notes that opposing school choice "is a bit like being asked to burn the American flag at a VFW meeting. You have every right to do it, but do you want to? After all, choice is a bedrock American value."[115] This is particularly true within the political culture of policy elites in Texas. Stefan Haag, Rex Peebles, and Gary Keith note that individualism is the single most important political value in Texas.[116] Throughout the state, the values of individualism and liberty are cherished above all others. In her study of school reform in Texas, Linda McNeil observes "the entrepreneurial presumption is that you take care of yourself and your own; if you are not successful right now, either your luck may turn any day, or you're not working hard enough."[117] This entrepreneurial, anti-bureaucratic spirit is common among policy elites in southern and western states.[118] Individualism is so pervasive that it is institutionalized in the operation of the state legislature, the executive leadership, and even incorporated into symbolic markers such as the state flag—a solitary star, hence the "Lone Star State."

All policy discourse is framed within this cultural context that carries "its own characteristic perspectives and ways of framing issues."[119] Arguments appeal to emotion, rather than rationality; anecdotes suffice for statistics.[120] In a dramatic example of the political manipulation of language, the Pennsylvania State Education Association, the state's largest teachers' union, posted a picture of former military dictator Augusto Pinochet on its website; Pinochet had instituted a voucher plan in Chile.[121] In another instance, a group of school district superintendents in Bucks County, Pennsylvania, urged state lawmakers to vote against a voucher proposal supported by Governor Tom Ridge, stating, "The current war in Kosovo is a graphic example of what happens in a society that separates its people and fosters elitism."[122] In these language games, words are carefully chosen, appropriate metaphors and analogies used, which appeal to popular cultural markers or referents. Patti Lather argues that this seemingly transparent use of language is not innocent.[123] She asserts, "The way we speak and write reflects the structures of power in our society."[124] According to Lather, "speech is part of a discursive system, a network of power that has material effects."[125] The material effects are manifested in the policies adopted (or rejected) through the policymaking process.

Drawing from Cleo Cherryholmes, language "determines what counts as true, important, or relevant."[126] For example, the illusion of choice "creates a pedagogy of entrapment that makes it undemocratic to argue against school choice. Thus school choice becomes part of a discourse that brooks no dissension or argument, for to argue against it is to deny democracy."[127] Choice, then, is viewed within the cultural referents of individualism, liberty, and democracy. To argue against choice, opponents are placed in the unenviable position of opposing cultural values widely held among policy elites. As Christopher Bosso suggests, policy change is difficult when it conflicts with strongly held public values in the political culture.[128] Conversely, policy change is easier when it reflects commonly held public values. In the context of this analysis, arguing against choice becomes un-Texan or un-American. As a result, few policymakers oppose school choice. Everyone, in fact, argues for *more* choice. The language game became how best to frame choice within the education policymaking context.

Another factor facilitating the Texas legislature's ultimate adoption of charter school legislation was that charter school rhetoric occupied a privileged position within the policy discourse. In the first language game, charter schools were characterized as freeing schools from the excesses of the state. In the second and third language games, charter schools were viewed as injecting "safe competition" and offering "safe choice" within the confines of the public school system. These findings are consistent with research suggesting that the major political advantage of charter schools is that they occupy something of a middle ground between the public education system as it is currently structured on the one hand and a voucher system on the other.[129] Some researchers describe charter schools as "experiments financed within the public school system."[130]

As Robert Bulman and David Kirp conclude in their analysis of the shifting politics of school choice, "The growing importance of choice in educational policy is partly explained by the ways advocates have been able to characterize their ideas, creating a more sympathetic picture by shifting the conversation away from an emphasis on market-based choice and toward an emphasis on equity-based choice," setting the stage for the adoption of charter schools as a preferred choice option.[131] Since charters are situated within the rhetoric of the public sphere, the scope of the conflict over charter schools is limited.[132] There was little debate as to whether charters should be granted to public schools in Texas. Questions were raised only about the number of charters to be granted. Utilizing the language of choice within the public sphere to structure discourse, supporters of charter schools were able to limit the scope of the conflict, facilitating passage of the legislation.

In the policy discourse over voucher proposals, however, supporters of voucher plans were unable to contain the scope of the conflict. Opponents were successful at expanding discourse within the second language game to encompass the issue of the traditional split between the public and private spheres—a distinction that many policy elites insisted on maintaining. Since vouchers directly threatened to eliminate the separation of the public and private spheres (or perhaps invert it, as Sheldon Wolin suggests many conservative policies attempt),[133] the scope of the conflict was expanded, the stakes of the language game raised. Drawing on the state's long history of unequal treatment of diverse populations, voucher opponents were able in the third language game to expand the scope of the conflict by raising issues of equity.

The suspicions of lobbyists such as Duggan and legislators such as Representative Turner and Senator Barrientos of the motives of many choice proponents are supported by several studies that conclude, as Michelle Fine did, that "the rhetoric of choice typically enters educational discourse when a privileged group seeks refuge from one public context and entrance into another, more elite context."[134] Since the scope of the policy discourse over vouchers was wider, conflict and controversy were unable to be contained. The resultant political conflict dimmed the prospects of passing voucher legislation. This conclusion lends support to research emphasizing the critical role of language in structuring policymaking and influencing the scope and outcome of conflict.[135]

RECONSIDERING THE ROLE OF LANGUAGE IN POLICY DISCOURSE

The first language game favored neither charter schools nor vouchers but rather action qua action, which ultimately privileged both reforms.

The second and third language games privileged charter schools over vouchers in the policy discourse, in large measure because the charter school reform did not threaten the traditional split between the public and private spheres. Charter schools were framed as "safe competition" and as "safe choice." Charter schools were viewed as less of a threat to widen the gap between the haves and have-nots within the public school system, privileging in the policy discourse the adoption of charter schools over vouchers.

From the perspective of the politics of language in shaping educational policymaking, a key area of additional research is whether policymakers are entering an era in which the distinction between the public and private spheres is disappearing or becoming inverted. Several states allow for-profit companies to operate charter schools. Does the blurring of the distinction between the public and private spheres, and the associated language games therein, increase the likelihood that voucher plans may pass? Will popular and elite acceptance of the arguments in language games played by charter school advocates facilitate the adoption of voucher plans? One state representative clearly believed it would. During an interview, he asserted:

> The analogy I use is that I think President Clinton and basically opponents of a market-based education system are treating charters kind of the way Gorbachev treated glasnost and perestroika in that, you know, the Soviet Union saw that they needed to change things, democracy was this big threat, they needed to introduce some reforms and glasnost and perestroika were seen as a way to forestall democracy and to forestall a free market and I think that the teachers' unions view charters as the same way, basically as a way to forestall choice. I also think that they are making the same, I think it is an enormous strategic blunder, the same as Gorbachev made an enormous strategic blunder in that by endorsing perestroika and glasnost he actually hastened the end of communism, and I think by endorsing charters, they are hastening the end of the education system as a monopoly, a government-run monopoly. It actually makes it easier to move onto vouchers. What it's done here in Texas, it destroys many of the arguments that are typically used against vouchers.

These issues need to be researched in greater depth since they have significant implications for the nature of public and private education in the United States. Additional research of the words and symbols policymakers use and how political culture affects the language games they play needs to be conducted to more fully understand how these games affect how policy options are viewed and alternatives evaluated.[136] As David Rochefort and Roger Cobb conclude, it is commonly said that actions speak louder than words. However, "in the world of politics and policymaking, this is not necessarily so" as words influence and even substitute for actions.[137]

Accordingly, longitudinal studies of the rhetoric used in language games need to be undertaken to more fully understand the subtle changes in politico-cultural attitudes and beliefs in language games over time. More attention must be devoted to analyzing how new metaphors are introduced successfully into language games, thereby shifting the parameters of policy discourse. As Prawat and Peterson point out, little is known about the rules governing language games, including how these games change over time.[138] Some philosophers such as Ludwig Wittgenstein argue that people "decide spontaneously" on new language games, although this is ultimately a simplistic and unsatisfactory explanation of a complex phenomenon.[139] Richard Rorty asserts that the introduction of new metaphors sparks changes in the language game (thus shifting the direction of policy discourse), although it is more likely that these "new" metaphors are simply old, recycled metaphors situated before a more receptive audience.[140] These metaphors arise out of a politico-cultural framework that shifts over time—for example, during periods of liberal or conservative dominance in state and national politics. Shifts in the mood of the electorate, or changes in socioeconomic conditions caused by an economic boom or recession, open rhetorical windows in language games, offering opportunities for key actors in a policy subsystem to use their position to advance particular policy agendas.

SUMMARY

In this chapter, we have explored the role of political culture in setting the context within which policymaking takes place. While Elazar's operationalization of political culture into three distinct political subcultures has been shown to be problematic, clearly policymakers are influenced by a more generalized political culture, as the analysis of their debates and rhetoric suggest. Political culture helps to set the parameters within which policy debates occur and this process is not neutral in its effects on policy consideration, debate, selection of alternatives, and policy adoption. Studies of policy rhetoric highlight the politics of language, manifested through the playing of language games, as participants struggle to dominate and control the policy discourse. The success of policymakers in either expanding or contracting the scope of rhetorical conflict plays a key role in shaping the policy process, as well as its eventual outcomes.

CHAPTER 3

INSTITUTIONAL DYNAMICS: THE POWER OF STRUCTURE

If political culture creates the sociocultural context of policymaking, then institutions provide the essential structure, the rules of the game, within which policymaking occurs. Institutional theory, developed in the fields of political science, history, sociology, and organizational studies, focuses not on actors as change agents but rather on the effects of institutions in shaping or mediating policy change. Consisting of numerous (and sometimes significantly) different approaches, this approach is generally known as the new institutionalism.[1] Although it encompasses a variety of perspectives, from rational choice to historical institutionalism, the core idea of the approach is the recognition that institutions of the state shape policy change.[2] Institutions play an integral role in defining individual, group, and societal identities and are, it is argued, much more than simple mirrors of social forces.[3] This perspective contrasts sharply with the behavioral point of view, whereby institutions are "portrayed simply as arenas within which political behavior, driven by more fundamental factors, occurs."[4] Institutional theory suggests that various institutional constraints play a key role in shaping policy choices and outcomes.[5]

Renewed attention to institutional variables "grew out of a critique of the behavioral emphasis" that "often obscured the enduring socioeconomic and political structures that mold behavior."[6] For example, interest group theories cannot account for why "interest groups with similar organizational characteristics (including measures of interest-group 'strength') and similar preferences" have different degrees of success in influencing policy outcomes, particularly in different national contexts.[7] The uneven success of interest groups in dominating the policy game, all other factors being nearly equal, is not well explained by traditional interest group theories.

The new institutionalism represents an attempt to focus on institutions as "ideological and structural devices for arranging the social, cultural, and political order."[8] For example, Peter Senge argues that "we must look into the underlying structures which shape individual actions and create the conditions where types of events become likely."[9] The structural context is critically important in policy formation and implementation because it provides the framework within which policy is crafted. This includes "various rules of procedure, including the constitution, statutes, prescribed jurisdictions, precedents, customary decision-making modes, and other legal requirements" as well as the mood of the public and the preferences of various politicians and interest group members.[10] For example, a federal system of governance facilitates institutional fragmentation, thereby providing multiple opportunities for interest groups to exert influence on policymaking.[11]

Institutions then become the object of analysis. Theda Skocpol asserts that "political activities, whether carried on by politicians or by social groups, [are] conditioned by the institutional configurations of governments and political party systems."[12] Therefore, the outcome of politics is mediated by the institutional setting within which policymaking occurs.[13] This approach explicitly links political preferences to institutional processes. Previous research assumed "that class, geography, climate, ethnicity, language, culture, economic conditions, demography, technology, ideology, and religion all affect politics but are not significantly affected by politics."[14] However, drawing from Stephen Krasner, the "preferences of public officials are constrained by the administrative apparatus, legal order, and enduring beliefs" of subsystem actors.[15]

One significant difference distinguishing institutional theory from other theories of policymaking and change lies in its treatment of preference formation and individual (and group) interest. James March and Johan Olsen argue that "if political preferences are molded through political experiences, or by political institutions, it is awkward to have a theory that presumes preferences are exogenous to the political process."[16] This suggests that public policy, as an outcome of institutional processes, shapes private preferences.[17] According to March and Olsen, "analysis of the effects of institutional variables on policy outcomes invites theoretical development of models of the ways in which interests and preferences develop within the context of institutional action, the ways reputations and expectations develop as a result of the outcomes of politics, and the ways in which the process of controlling purposive organizations produces unanticipated consequences and is tied to a symbolic system that evolves within an institution."[18]

Within institutional theory itself, significant differences exist regarding preference formation. For example, rational choice approaches treat preference formation as exogenous to the political process, while historical institutionalism considers preference formation as endogenous to politics.[19] Historical institutional approaches have greater applicability because they more accurately describe the complexities of politics. As Kathleen Thelen and Sven Steinmo argue, "Taking preference formation as problematical rather than given, it then also follows that alliance formation is more than a lining up of groups with compatible (preexisting and unambiguous) self-interests. Where groups have multiple often conflicting interests, it is necessary to examine the political processes out of which particular coalitions are formed."[20] Given the limits of the rational choice approach and its inability to capture the complexities of political behavior, the rest of this analysis focuses on historical institutionalism.

HISTORICAL INSTITUTIONALISM

Historical institutionalists posit that institutional structures affect individual political behavior and shape policy change.[21] Krasner asserts that "the ability of a political leader to carry out a policy is critically determined by the authoritative institutional resources and arrangements existing within a given political system."[22] According to James Cibulka, "Institutions both shape and constrain the choices actors make. They do this in numerous ways: by creating symbols and legitimating myths about the institution, by structuring the environment, by creating structures and processes for addressing goals and mediating conflicts, and so on."[23] Thus, both the goals and strategies political agents pursue "are shaped by the institutional context."[24] Internal institutional processes affect the distribution of power within organizational systems.[25] Decisions are affected by the power and position of those who make them.[26] Therefore, the state plays a key role in mediating competition among political actors and interest groups and it cannot be assumed to do so in a neutral manner.[27] As a result, institutions are much more than mere arenas for interest group conflict; they exert independent effects on policy outcomes.[28] Put simply, "the organization of political life makes a difference," as demonstrated in the intense political battles over charter schools and vouchers.[29]

E. E. Schattschneider long ago recognized the institutional bias of structure when he asserted that "the function of institutions is to channel conflict; institutions do not treat all forms of conflict impartially."[30] He went on to argue that "all forms of political organization have a bias in favor of the exploitation of some kinds of conflict and the suppression of others because organization is the mobilization of bias. Some issues are organized into politics while others are organized out."[31] This led Theda Skocpol, a major proponent of neo-institutional approaches to studying politics and policy outcomes, to conclude that "the overall structure of political institutions (including multiple points of access) provides access and leverage to some groups and alliances, thus encouraging and rewarding their efforts to shape government policies, while simultaneously denying access and leverage to other groups and alliances."[32]

Although it would be erroneous to argue that the state determines the outcome of conflict since it is one of several factors or inputs into the policymaking process, the institutional structure of the state does provide the channels through which policymaking and political conflict must flow. As Douglas Abrams suggests, "Structure is not an independent determinant of policy outcomes, but a conditioner of the political process by which those policy outcomes are achieved. It contributes to the ways in which issues are formulated, and the shape and quality of the political dynamics surrounding them."[33] Thus, the "state is not simply a neutral converter of societal demands" but rather an active participant in shaping the process of policy change.[34]

BRINGING THE STATE BACK IN

Given the utility of bringing the state back into models of policy change,[35] it might be useful to more clearly define what exactly is meant by the "state." According to Peter Hall, the state may be "broadly understood as the executive, legislative, and judicial apparatus of the nation."[36] Since each branch or division of the state asserts jurisdiction over specific areas, sovereignty claims are legitimated through the concentration of government power.[37] Thomas James utilizes an even broader definition of the state, "The state can be conceptualized in its broader meaning as the totality of public authority that constitutes the political system at all levels, or more concretely as American state governments, which have primary constitutional authority over formal schooling in the United States."[38] James noted a "growing state presence in shaping educational policy,"[39] brought about by an expansion in state capacity and "the use of law as an instrument for managing local schools."[40] This expansion in state capacity extends well beyond education to all areas of policy and is evident in increased state-level policy activism.[41]

Although historical institutionalism places the state at the center of analysis, it does not ignore the crucial role of interest groups and coalitions in shaping policy change vis-à-vis institutional processes. For example, in one of the earliest analyses using historical institutionalism, Peter Katzenstein notes that "the governing coalitions of social forces in each of the advanced industrial states find their institutional expression in distinct policy networks which link the public and private sector."[42] These coalitions and networks are central agents of policy change, utilizing the institutional structure to achieve their policy objectives. These objectives, in turn, are "shaped largely by the ideological outlook and material interests of the ruling coalition. Such coalitions combine elements of the dominant social classes with political power-brokers finding their institutional expression in the party system."[43]

Perhaps the greatest utility of the new institutionalism as it applies to policymaking and change is that it reminds us that "policies, once enacted, restructure subsequent political processes."[44] David Robertson, a proponent of the new institutionalism, argues that "past decisions shape the institutional constraints and opportunities of later periods, including the present."[45] Thus, as politics create policies, policies may be said to remake politics.[46] These policies "affect the social identities, goals, and capabilities of groups that subsequently struggle or ally in politics."[47]

The new institutionalism encourages researchers to look beyond existing political alliances and incorporate a historical dimension into their analyses.[48] This is necessary because "decisionmakers always draw on past experience, whether conscious of doing so or not."[49] For example, few educational policies are new. Most have long political histories, and many reforms are connected or linked, in one way or another, to previous policies. These political histories, the battles and outcomes, consequences and residual effects, play a role in shaping the outcome of current policy debates. Such a historical approach "furnishes a span of institutional time embracing a wide variety of changing conditions and variables,"[50] enabling researchers and policymakers to better understand the development of policy over time.

Margaret Weir suggests that researchers should look for connections among policies over time and view policy innovations as part of a policy sequence.[51] Applying this logic to charter schools and vouchers, one might view the school choice movement as part of a general political trend toward neo-conservatism and market-based educational reforms. Furthermore, researchers might examine the relationship between school choice initiatives (such as charter schools and vouchers) to see how some initiatives facilitate or impede the adoption of others—which is what this book is all about. The argument is to situate school choice within the context of larger politico-historical forces and to view the policy change from a historical, contextual lens.

However, great caution must be exercised when viewing policy as part of a policy sequence or as policy evolution—processes that imply a rational, logical, and orderly development from one policy iteration to another. While policies may be refined and improved over time, as I suggest has been the case with some educational accountability policies (see chapter 5), it does not follow that all policy change or even the majority of it (whether the area be educational, environmental, welfare reform, etc.) represents an improvement or an evolution over previous policy. This argument is detailed in the discussion of the role of organizational or policy learning in school choice contained in chapter 5.

The failure to adequately consider the effects of history and context leads to inadequate conceptualizations of social phenomenon.[52] Thus, one of the greatest strengths of the historical institutionalist approach to policymaking is that it seeks to develop theory at the middle range, allowing researchers to "integrate an understanding of general patterns of political history with an explanation of the contingent nature of political and economic development, and especially the role of political agency, conflict, and choice, in shaping that development."[53] Thelen and Steinmo argue that historical institutionalism:

> structures the *explanation* of political phenomena by providing a perspective for identifying how these different variables relate to one another. Thus, by placing the structuring factors at the center of the analysis, an institutional approach allows the theorist to capture the complexity of real political situations, but not at the expense of theoretical clarity. One of the great attractions and strengths of this approach is in how it strikes this balance between necessary complexity and desirable parsimony.[54]

To examine the impact of institutions on state policymaking in general and school choice in particular, we must first delineate the dimensions of state capacity for reform.

DEVELOPING INSTITUTIONAL CAPACITY FOR REFORM

David Robertson suggests that the policymaking capacity of the state can be gauged on three dimensions: "First, the formal boundaries of legitimate government intervention (that is, what it is permissible for government to do); second, government's fiscal ability (the ability to raise revenues and fund policy initiatives); and third, the professionalism and expertise of legislators and public administrators."[55] The formal boundaries of the state are generally defined by the constitution, which, in the case of state policymaking, can vary considerably from one state to another. Some state constitutions, like New York's, are much more expansive of state power, while others, like Texas, are "weak" and offer limited government opportunity for activism. The fiscal ability of the state is also constrained considerably by the constitution, as well as by economic conditions and resources (both people and material). Finally, significant differences have been found in the state legislatures in the degree of professionalism and expertise among legislators. Seven state constitutions mandate brief, biennial sessions. Some state legislatures are highly professional—with annual sessions, good pay, large staffs, and up-to-date technology—while others rarely meet, are poorly paid, and inadequately staffed. All of these factors affect the institutional capacity of states to engage in reform initiatives.

Other factors associated with the expansion of state capacity include the movement toward single-member districts, unlimited sessions, presession organization, uniform rules, growth of professional staff, and legislative management improvements such as electronic voting, bill-introduction deadlines, and the installation of "electronic data-processing equipment to track bills and reveal their contents."[56] David Robertson and Dennis Judd note that "over time, Congress and state legislatures, the president and state governors, and bureaucracies at all levels of government have grown larger, and more professional,"[57] enabling actors within state government to be increasingly active in shaping education policy and initiating policy change.

As a result of increased state capacity, "we have mounting evidence that, indeed, the states did take charge, with all fifty making important changes in their schools."[58] Joseph Murphy observes that there "has been a dramatic increase in the capacity of state governments to engage in educational issues."[59] State legislatures and their staffs have become more professionalized, and judicial interventions and gubernatorial initiatives have increased, enabling state governments to "use the means at their disposal to influence schooling at the local level" to a degree unprecedented in U.S. history.[60]

Coupled with the institutional fragmentation of a federal system of governance in which education is primarily a state (and local) responsibility, the institutional structure of the state gives well-positioned participants, such as the chairs of education committees and their finance subcommittees, the power to structure the agenda-setting process.[61] The committee structure provides considerable institutional influence over policy outcomes.[62] Thus, the actions of political agents are, according to Stephen Skowronek, "mediated by the institutional and political arrangements that define their positions and support their prerogatives within the state apparatus."[63] Tim Mazzoni, one of a handful of educational researchers who have devoted their careers to studying the politics of education policymaking at the state level, observed that "all other would-be initiators must deal with the preferences, power, and personalities of these key lawmakers,"[64] allowing these participants to set the terms or parameters of debate and, ultimately, to play a decisive role in determining the outcome of policymaking.[65]

INSTITUTIONAL DYNAMICS AND THE POLITICS OF SCHOOL CHOICE

Within the arena of school choice, institutional and organizational structures, rules, and procedures play a key role in shaping legislative outcomes. To assess the effects of institutions on policymaking, we must understand both the power of structure and the structure of power. Institutions provide policy elites with enormous influence over the policymaking process; elites, in turn, utilize this power to reinforce, maintain, and preserve their dominance of the policy process.

A growing body of evidence suggests that the institutional structure of governance in the United States significantly affects battles over school choice, both in Washington and in the fifty state legislatures, usually serving as a substantial constraint on policy change. For example, in 1997, a voucher bill targeted at children from low-income families in the District of Columbia passed both the House and Senate, only to be vetoed by President Clinton.[66] That same year, Clinton vetoed a plan to provide families with educational savings accounts that could be used to "pay for tutoring, school fees, home computers, preparatory test courses for college, or private/parochial school tuition for grades K-12."[67] To see this process in action at the state level, we turn to specific instances of institutional effects on school choice.

State Constitutional Constraints

State constitutional strictures place significant constraints on efforts to pass school choice legislation. State constitutions do much more than merely provide structure, the "rules of the game," to politics; they create conditions that privilege mild or moderate forms of choice such as charter schools over more radical forms of choice such as vouchers. In their analysis of efforts to expand school choice in Michigan, Hubert Morken and Jo Renee Formicola found that the state's constitution was a major impediment to any voucher or tuition tax credit plan, since it forbade the use of public money in private or parochial schools. The authors conclude that these measures could only be achieved through the initiative process, since the votes necessary to amend the state constitution "are never going to be there."[68] In Michigan, voucher plans have met a brick wall in the form of the state constitution, which "includes an airtight prohibition against public funding for nonpublic schools."[69]

With respect to the substantial constraints imposed by state constitutions, voucher supporters were given a significant boost with the recent U.S. Supreme Court ruling in *Zelman v. Simmons-Harris*[70] upholding the constitutionality of Ohio's school choice program, in which low-income parents in Cleveland can use public funds to enroll their children in other public, private, or religious schools.[71] In a majority opinion, the Court determined that the program's provision for use of education vouchers in religious schools was "one aspect of a broader undertaking to assist poor children in failed schools, not as an endorsement of religious schooling in general."[72] The Supreme Court's ruling, however, does not clear all the constitutional hurdles posed by voucher plans. Thirty-seven state constitutions specifically prohibit state aid to religious schools.[73] At the very least, such constitutional strictures will delay the passage and implementation of voucher initiatives across the country, with several court challenges underway.

The judicial system has been used by both proponents and opponents of charter schools as a venue for achieving policy change. Scores of lawsuits have been brought by teachers' unions, local school boards, school board associations, and others challenging states' charter school laws, albeit with little success. Generally, suits brought by or on behalf of charter school operators seeking to overturn administrative obstacles imposed by superintendents and school boards have met with greater success.[74]

In Texas, the institutional structure favored the adoption of charter schools over vouchers. For example, several respondents indicated that the state constitution has slowed the push toward vouchers. Opponents asserted that they would use the legal process to block any voucher bill passed by the legislature. In an interview, a senior official with the Texas Elementary Principals and Supervisors Association (TEPSA) stated "the ACLU has already said they would sue." Another member of a teacher's group said "the minute a pilot program passes, there will be a lawsuit filed." A senior official in the state education agency gave his opinion that "I think with vouchers you are going to have a real significant constitutional problem with the use of state monies in private school systems, in parochial schools, church-related schools. I think there are at least three references in the Texas state constitution about using state funds to support church-affiliated schools." In fact, references in several state constitutions to the separation of church and state are being used by members of anti-voucher coalitions throughout the country as a line of defense against voucher plans.

Richard Kouri, president of the Texas State Teachers Association (TSTA), compared the coming political conflict over vouchers to the battle over school finance, stating, "It's a lawsuit that will make school finance look small," a potentially powerful threat given the decades-long legal battles over school finance fought in state courts throughout the country.[75] The tremendous amount of political turbulence surrounding the lengthy and expensive battle over school finance has acted as a deterrent discouraging state legislatures from adopting voucher plans, although the *Zelman* decision will undoubtedly spur renewed interest in voucher plans. In fact, one member of the anti-voucher coalition hoped that the amount of political conflict generated by a voucher plan might force the legislature to drop the plan and adopt less controversial reforms.

Several members of the anti-voucher coalition said they would use the legal structure of the state to block the implementation of voucher plans, a strategy used by anti-voucher forces in other states. In Pennsylvania, Governor Tom Ridge predicted that legal challenges could tie up implementation of vouchers "for two or three years."[76] Members of pro-voucher coalitions must contend with these barriers when developing voucher plans in state legislatures. Whatever the outcome of these battles, it is clear that the *Zelman* decision will not lead to the rapid adoption of vouchers by state legislatures. It will take years of legal challenges before this issue is settled by the courts. With respect to vouchers, state constitutions play a major role in shaping the content of voucher plans as they wind their way through state legislatures.

In addition to the constitutional strictures noted above, several state constitutions, such as that of Texas, do not provide for the initiative or referendum process, which has been used by voucher supporters to place statewide voucher proposals on the ballot.[77] In the twenty-four states that allow for either the initiative or referendum process, voucher supporters have used these processes to circumvent intransigent anti-voucher state legislatures and governors, albeit with little success.[78] Voucher plans have been rejected by wide margins by voters in California, Colorado, Michigan, and Oregon, to name only a few. In 1993, voters in California rejected Proposition 174 by a 2-1 margin. Seven years later, California voters rejected Proposition 38, which would have given parents who already send their children to private schools a $4,000 voucher per child; the measure failed by nearly 3-1. Michigan voters rejected a voucher initiative in 1978 and again in 2000; the 2000 measure, Proposal 1, would have provided private school tuition vouchers for children attending schools in low-performing public school districts.[79]

Length and Frequency of Legislative Sessions

The length of the legislative session and the frequency with which the legislature meets also affect school choice initiatives. Research suggests that "no single factor has a greater effect on the legislative environment than the constitutional restriction on the length of [the legislative] session."[80] Texas is one of only seven states with biennial sessions, and it is by far the largest (both in size and population) and most complex state with biennial sessions. The sessions only last for 140 days every biennium. In effect, the legislature meets for five months, and then disbands for the next nineteen months. Coupled with extremely low legislator pay (set in the state constitution at $7,200 per year), and the massive number of bills introduced each session (at least 5,000 bills per session), the Texas Legislature reflects the founders' intent to create a restricted, limited government led by a part-time collection of "citizen politicians."[81] In addition, unlike in many other state legislatures, only the governor can call the legislature into special session to handle any unfinished business at the end of the session.

In Texas, the shortness of the legislative session has been detrimental to the coalition supporting vouchers, although it has facilitated the passage of charter school legislation and helped groups pushing to create charter schools and later expand the cap on the number of such schools. During the 1995 legislative session, after three days of floor debate and 400 proposed amendments, the House passed a comprehensive plan to reform public education in Texas (S.B. 1).[82] The debates surrounding charter schools and vouchers occurred amidst debates over hundreds of other provisions in the bill, including curriculum, textbooks, discipline, teacher and administrator preparation, accountability, school finance, and special education.[83] Senator Bill Ratliff (R), then-chair of the Senate Education Committee, said of proposed revisions to the code, "There is certainly something in this bill to offend everyone. Any time you're going to try to totally rewrite something like the public education code, you have to step on a lot of toes or you're not ever going to get anything done."[84]

In their analysis of modifications to special education policies included in S.B. 1, researchers Virginia Baxt and Liane Brouillette quote the chair of the Senate Education Committee (speaking about S.B. 1), "I think one of the interesting things about Senate Bill 1 is that there are hundreds of things in S.B. 1 that might not have ever been passed had they been stand-alone pieces of legislation... Generally, big bills are hard to pass. But when it becomes this massive, this huge overhaul, a legislator is able to say, 'Well, I don't like this piece, or that piece, but we need to overhaul.' "[85] In the case of debate over various school choice proposals to be included in S.B. 1, a compromise was made and charter schools were adopted, due in large measure to the fact that they were much less controversial than other choice proposals such as vouchers.

In 1997, much of the legislature's time on education-related issues was spent on tax and school finance reform measures. According to materials provided by the Texas Classroom Teachers Association, this focus contributed to the reduced consideration of other measures.[86] Given these rather severe time constraints, measures that generate much controversy, such as vouchers, are much less likely to win legislative approval. A member of the pro-voucher coalition put it this way: "You'll find that [the] middle of the road, and it spans the political spectrum, Democrats and Republicans, tend to want to do something to help children and they have seen the ideas that you see in traditional public schools as a failure and so many people... are now starting to hop on to the charter school bandwagon because they see it as something that is not quite [as] controversial [as vouchers]." The shortness of the legislative session also affects the ability of interest groups to influence legislation. A senior staff member believes part of the reason the cap on charter schools was increased was "just confusion at the end of the session. The last day of the session. I hate to say no one was paying attention but it kind of got through because the Governor was pushing it. A lot of people didn't want to kill the governor's primary legislation. There was a lot going on and there wasn't time to get an organized opposition to it."

In the House, a number of pro-voucher bills were introduced to the House Education Committee and to the House Select Committee on Revenue and Public Education Funding. The Select Committee, where most of the bills were assigned, was preoccupied with property tax reform proposals for most of the short legislative session. Because of the time crunch, none of the voucher schemes was acted on.[87] In 2001, the Texas Legislature was dominated by concerns over redistricting, health care, and school finance. Redistricting is a highly contentious process that makes "legislators so angry with each other that compromise on almost anything" becomes nearly impossible.[88] The only significant school choice legislation produced by the state legislature during the 2001 legislative session was H.B. 6—a bill that strengthened state oversight and accountability of charter schools.

The shortness of legislative sessions in Texas places a premium on interest groups being well organized, proactive, and initiating lobbying efforts prior to the start of the session. The Texas legislature is seldom in session. Not surprisingly, this produces rather high legislator turnover, as few legislators make it a career. Approximately one-third of House and Senate members in the 2003 Legislative Session are freshmen, non-incumbent officials. Service in the legislature remains very much a part-time job, which is exactly what the framers of the Constitution of 1876 intended. The high turnover of state legislators, resulting in relative unfamiliarity with the legislative process, makes these legislators more open to the influence of lobbyists.[89] Texas has been characterized as having a strong pressure group system.[90] The characteristics traditionally associated with strong interest group systems include short legislative sessions, frequent turnover, amateur status of state politics, and the part-time nature of the job, all of which describe the Texas legislature.[91] According to a study by the Citizens Conference on State Legislatures, "A legislature that meets only for a few months every other year cannot dig very deeply into the whys and wherefores, the facts and figures, the social and economic implications, of the legislation that comes before it. Nor can it very readily accumulate information, or undertake any searching analyses or investigations, on any continuous basis if its life is altogether confined to those few, fleeting months."[92] Despite all the recent attention paid to vouchers, it must be remembered that the reform remains only one of a number of education issues facing state legislatures, including the significant implementation challenges presented in the federal No Child Left Behind Act of 2001.

Furthermore, faced with a balky economy and increasingly tight fiscal constraints, state lawmakers may find it difficult to maintain current levels of funding for education, much less initiate new programs.[93] According to Todd Ziebarth, policy analyst with the Education Commission of the States, "Most states are dealing with deficits, and their energy is focused on balancing budgets."[94] Economic constraints significantly affect the ability and willingness of state policymakers to undertake education reform initiatives.[95] "Rapidly rising health care costs, a weak economy, and falling revenues," coupled with increasing federal mandates, portend a budget crisis in the states that will necessitate tough spending decisions.[96]

Such pressing imperatives may limit the amount of time devoted to voucher initiatives in forthcoming legislative sessions, despite the *Zelman* decision, particularly given the degree of conflict such proposals generate. Similarly, in the U.S. Congress, a backlog of pressing legislation, "including 13 must-pass spending bills, a prescription-drug bill, and a plan to create a homeland security department", coupled with military conflict in Iraq, concerns over North Korea, and domestic terrorism leaves little room for controversial, time-consuming debates over voucher proposals.[97] Thus, when assessing the prospects of voucher legislation, we must be careful to examine each state's legislative picture in its totality—economic concerns, constitutional impediments, legislative priorities, history, and political culture, to name but a few.

Sometimes, support for and opposition to school choice initiatives is tied to other education reforms, such as mayoral control of schools. For example, in 1996, Ohio House Education Committee Chair Michael Fox, a Republican, made it known that state legislators were ready to turn over control of the Cleveland public schools to Mayor Michael White—a move which the Cleveland Teachers Union opposed. Fox suggested that if union leaders softened their opposition to school choice, then he "might just have ammunition enough to slow the mayoral bill's momentum."[98] As this example demonstrates, "other policies under consideration at the same time as charter schools can distract opponents and influence the compromises that policymakers are willing to make."[99]

Institutional Power of Governors

Governors play a key, although not determinant, role in promoting school choice. In an interview, a senior official with the pro-voucher coalition in Texas stated, "the Governor himself shapes the debate (over school choice). The Governor [Bush] called Republicans in this session that were wavering on vouchers and talked with them." However, the formal powers of governors vary considerably from state to state. For example, compared to the governors of most other states, the governor of Texas has little formal institutional power. Historically, in terms of formal powers, the governor of Texas is one of the weakest governors in the nation.[100]

This condition reflects the framers' intent, under the Constitution of 1876, to create a weak governor subordinate to the wishes of the legislature.[101] Unlike the governors of other large states, the governor of Texas has no cabinet and severely limited powers of appointment.[102] Although Texas has approximately 200 state agencies, the governor has power of appointment over only 12 of those agencies. The only major statewide official appointed by the governor is the secretary of state. All other statewide officials are elected independently, a system known as a fragmented or plural executive. Even the lieutenant governor is elected independently of the governor. In Texas's political history, the lieutenant governor traditionally wields much greater formal power than the governor.

Former state senator Kent Caperton, who served in the state House from 1981 to 1990, stated that "the framers of the Texas Constitution were very careful to set up the respective powers so that it was impossible then and now for a governor to dictate to the Legislature what to do or not to do."[103] Billy Clayton, House speaker from 1975 to 1982, explained that "a governor who tries to run roughshod over the Legislature will 'wake up and find retaliation.'"[104] Thus, while much attention has been focused on the increasing importance of governors as policy entrepreneurs in the school choice movement, their powers are constrained by a number of forces, including state constitutions, economic forces, and the necessity of satisfying diverse electoral constituencies.[105] For example, despite making vouchers one of his top education priorities when he took office in 1995, Pennsylvania Governor Tom Ridge, like Texas Governor George W. Bush, failed three times in five years to push a voucher bill through the state legislature. Although gubernatorial interest and activism in shaping state education policy have increased in recent years, these examples highlight definite limits on the ability of governors, even popular governors, to remake the educational system in their own image.

Much of a governor's power lies in their ability to build coalitions. Max Sherman, then-dean of the Lyndon B. Johnson School of Public Affairs at the University of Texas, stated that "it's [the real power of the governor] in the abilities to persuade, lead, and work cooperatively with people in positions of leadership in the Senate and the House."[106] Governor Bush clearly understood his role in policymaking when he stated, "my dream is to rise above partisan politics.

And that's why I'm going to get along with [Lieutenant] Governor Bullock and Speaker Laney."[107] Bullock replied, "We proved last session that nonpartisanship is the key to getting anything done. I don't expect anything any different this time. Why argue with success?"[108] Former state Senator Ike Harris, who spent nearly 30 years in the legislature, noted that "the combative approach may be what gets you on the front page, but it's better for a governor to work out any problems with legislative leadership in a friendly, quiet manner."[109] Thus, although Governor Bush favored a pilot program for vouchers, he had little formal power to push such a bill through the legislature.

Institutional Power of Legislative Leadership

In state legislatures, key leaders such as the Speaker of the House and Senate President exert substantial control over legislation—determining which bills rise to the top of legislative agendas and which bills languish forever in committee, never to see the light of day. The rules and procedures of many state legislatures give greater power to these key leaders than to leaders in the U.S. Congress, "where standing committees and numerous subcommittees have balkanized the national legislative process."[110] It is not uncommon for state legislatures to adopt rules centralizing authority and influence in the hands of relatively few leaders. In Ohio, for example, the Speaker of the House and President of the Senate chair their respective Rules Committees, which gives these legislators enormous control over the legislative process.[111]

The following example highlights the power of these legislative leaders. In 1997, the Ohio General Assembly passed legislation appropriating $4.5 million for charter schools in Toledo. However, Senate President Richard Finan, a Republican, hinted that the provision might be removed from the bill "to punish a Republican House member from Toledo who opposed his [Finan's] efforts to exempt school construction projects from prevailing wage requirements."[112] The representative, Sally Perz (R-Toledo), had opposed the exemption, which was one of President Finan's favorite bills. A few days later, Senate Republicans removed the charter school funding, one of Representative Perz's pet projects, from the proposed state budget.[113] This example highlights the power of legislative leaders both in shaping the content and in controlling the flow of legislation. On another occasion, "In a not-so-subtle reminder to Senate Democrats about who controls the fate of their bills, Mr. Finan delayed a floor vote on three Democratic measures, even though they passed committee with bipartisan support."[114]

In the state House, two key leadership positions carry enormous institutional power, Speaker of the House and chair of the Calendars Committee. A member of a pro-voucher group recognized the important role of the Speaker of the House in affecting school choice legislation in Texas. He stated, "the Speaker of the House has a tremendous influence on vouchers [he opposed them]." A senior committee staff member in the Senate said members of the pro-voucher coalition made a serious tactical error by getting involved in the fight to defeat House Speaker Pete Laney. Pro-voucher advocates had become displeased with Speaker Laney's lack of passionate advocacy for vouchers. The staff member said in an interview, "You pissed off Pete Laney. And Laney is not a liberal. Laney is as conservative as you can get. Laney is one of these Speakers that says, 'Let the will of the House decide' but getting out there and recruiting somebody and pouring money into his opponent and trying to defeat him and going around telling people he is the devil is just stupid politically." A spokesperson for a teachers' group agreed, stating "I know Speaker Laney back in '95, he actually walked the floor and said, 'Vote with me on this' (against vouchers). You pretty much go along with him or you'll never get a committee appointment worth having."

Under the Texas state constitution, the Speaker has the power to appoint the chair and vice chairs of all standing committees and assign bills to committee. This gives the Speaker enormous influence over legislation—power that Speaker Laney used to block voucher bills from House passage. In 2003, with the Republican takeover of the House, the Texas House of Representatives is led by Republican Tom Craddick, a voucher supporter. Craddick's reign as Speaker marks the first time since 1871 that a Republican has occupied the top leadership position in the House. Republicans have not been the majority party in the House in over 130 years. As a result, the probability of voucher legislation passing through the House of Representatives is better now than it has ever been. A pro-voucher Speaker could play a major role in shepherding voucher legislation through that chamber, although it is by no means certain that vouchers are a done deal in Texas, as will be discussed in detail later in this chapter and throughout this book.

As in the House, the views of key leaders in the Senate affect the chances for passing legislation. A state representative thought the chances of passing a voucher bill in Texas were not good. He stated,

"In '97, they (supporters of vouchers) didn't even get a vote in the Senate. I think that now, and it's going to become more so as time goes by, it's going to be more difficult to get it done in the Senate." When asked why he believed this, he responded "because you have some Republicans who are in leadership positions who really don't believe much in vouchers. And that's Ratliff, Teel Bivins, and Duncan." He noted that Duncan is firmly against vouchers.

The position taken by the lieutenant governor carries considerable weight. In Texas politics, the lieutenant governor has wielded the most political power in the state, even more than the governor. According to a study by the Citizens Conference on State Legislatures, the lieutenant governor of Texas wields more real legislative power than the lieutenant governors of most states.[115] Bill Miller, a political consultant who has worked for both Republicans and Democrats, said of legendary Lieutenant Governor Bob Bullock, "he's the Don. He is a metaphorical death knell if he is opposed to you."[116] A senior staff member believes the lieutenant governor played a pivotal role in the adoption of charter schools in Texas. He stated,

> I tell you what was really an impetus for it [choice] is . . . the lieutenant governor [Bullock] authorized his staff to represent at the committee hearings through testimony that he strongly favored the idea of a charter school or a charter school concept or charter school public policy. He felt like we needed to give more opportunities for parents to make choices about what kind of education they wanted their children to have and that this was particularly true for minority children who often are trapped in schools that evidence, academic performance, below a standard that would be acceptable. Given how the legislature is structured, which is, I mean, it really is kind of run as a dictatorship with the lieutenant governor being the dictator. When the lieutenant governor says, "This is what I want," then those members in the Senate, to the extent that they can, will try to make sure he gets it.

Although Bullock had at one time supported a pilot program for vouchers, he did not actively push voucher plans in either the 1993, 1995, or 1997 legislative sessions.[117] This lack of strong advocacy played a key role in the defeat of vouchers in each of those sessions. In the 2003 legislative session, the Lieutenant Governor is Republican Bill Ratliff, a moderately conservative legislator who supports a pilot program for vouchers. Coupled with a pro-voucher Speaker in the House, the chances of passing voucher legislation are reasonably good.

Institutional Structure, Committees, and Choice

The division of the work of state legislatures into committees also influences the flow of legislative proposals. The committee structure allows highly controversial legislation to be "avoided or killed in committee."[118] Appropriations committees are useful tools for blocking controversial legislation. For example, in Pennsylvania, the appropriations committee "acts like a 'super committee,' bottling up legislative proposals for both substantive and fiscal reasons."[119] The Speaker also appoints the chair and majority members of the all-powerful House Calendars Committee. The Calendars Committee schedules bills for debate by the House, determines the length of time allotted for debate, and whether amendments can be added on the House floor.[120] In an interview, one senior House staffer put it bluntly, stating Calendars "is where you kill bills."

A state representative emphasized the crucial role of the institutional structure of the House in affecting the outcome of school choice legislation in Texas. He explained:

> The way the House works, you have to get through a committee, and the House Education Committee has traditionally been, leaned toward liberal Democrats, not just Democrats but liberal Democrats, and so it's almost impossible to get out of that committee. If you do get out of that committee, then you have got to go to the Calendars Committee which sets everything up for a vote. The Calendars Committee can decide whether or not they want to set your bill for a vote, even after it has gone through committee. Traditionally, that has leaned toward being liberal Democrat and traditionally they pay some attention to the will of the Speaker of the House and the will of the Speaker of the House, Pete Laney, has been against vouchers.

In 1999, Texas Lieutenant Governor Rick Perry stripped the Senate Education Committee of two of its staunchest voucher opponents, and assigned the chair to voucher advocate Teel Bivins.[121] During the 1999 legislative session, Bivins introduced S.B. 10, a five-year voucher pilot program in the state's six largest counties. Bivins estimated that 143,000 students would be eligible for the program.[122] Three weeks later, the bill was voted out of committee by a 5-4 vote along party lines, although it failed to be considered by the full Senate.[123]

Blocker Bills and Supermajorities

Often, unique rules and operating procedures in state legislatures exert a significant impact on proposals for school choice. Several

Table 3.1 Texas election results, 1982–2003

	1982	1988	1994	2003
House				
Democrats	114	93	89	62
Republicans	36	57	61	88
Senate				
Democrats	26	23	17	12
Republicans	5	8	14	19

Sources: *Statistical Abstract of the United States, 1982, 1989, 1994* (Washington, D.C.: U.S. Bureau of the Census, 1982, 1989, 1994); *Texas House Directory*, 1997; www.ncsl.org/programs/legman/elect/statevote2002.htm; www.house.state.tx.us; www.senate.state.tx.us.

proposed voucher bills have been defeated in the Texas Senate in the last four legislative sessions, including the 1997 legislative session where Republicans occupied a majority for the first time since Reconstruction (see table 3.1). Given strong Republican support for vouchers, this appears to be an anomaly. Interviews and archival research reveal the importance of institutional rules in defeating voucher plans, as demonstrated in the following example.

During the 1997 legislative session, in a 6-4 vote divided along party lines, the Senate Education Committee approved a scaled-back voucher program.[124] Oddly enough, members of the anti-voucher coalition soon proclaimed victory. A member of the Coalition for Public Schools stated, "we have 11 votes—enough to keep it [a voucher plan] from coming up in the Senate."[125] The Senate operates under a procedural rule called the "Two-Thirds Rule." At least two-thirds of the Senate must vote to bring a bill out of committee to the floor for a vote. Since the state Senate has only 31 members, this "Two-Thirds Rule" allows a minority of 11 senators to prevent any legislation from coming to the floor for debate, regardless of the strength of its support when coming out of committee. A senior staff member familiar with Senate rules and procedures explained how the Two-Thirds Rule works:

> At the beginning of the session, there is a bill filed. And one of the rules of the Senate is you have to take bills up in the order that they are filed or the order that they get to the Senate floor. And in order to suspend the rules and take up a bill out of order, you have to have a two-thirds majority. Well, this bill that's filed or gets to the floor of the Senate first is called the blocker bill. Nobody ever requests, whoever the sponsor

> is, it's somebody that the lieutenant governor trusts, never requests that that bill come up for debate. So every bill that comes up on the floor of the Senate is coming up out of order. And you have to have two-thirds of the senators to bring it up for debate. That's done as a housekeeping measure because there are so many, so many bills and unless they have strong support, why waste the Senate's time? We don't have much time to deal with things around here.

Time is at a premium due to the brevity of the legislative session. The vast number of pressing issues that legislators must address leaves little time for consideration of highly contentious proposals.

A senior senator noted that the Two-Thirds Rule blocks much legislation from being debated by the Senate. He stated, "you are going to have to have a two-thirds vote to bring anything to the floor. And so, as a result, 11 members of the Senate can block any bill." In effect, passing legislation through the Texas Senate requires a two-thirds majority (in effect, a supermajority), followed by a simple majority once the bill is on the Senate floor. Texas is one of few states in the nation to have such a procedural rule, explaining why votes in the state senate are seldom close—the overwhelming majority of legislation passed by the senate receives few dissenting votes. In their analysis of government capabilities in the United States and abroad, Kent Weaver and Bert Rockman observe that, "Voting rules that require supermajorities for legislative action may affect government capabilities" and legislative outcomes.[126]

The Two-Thirds Rule has had a significant impact on school choice initiatives in Texas, effectively blocking voucher legislation. After all, Texas is a very conservative state where freedom, individualism, and choice are bedrock cultural values. Accordingly, like Arizona, one would expect that voucher proposals would receive widespread legislative support. However, in 1999, 11 senators announced their intention to block a voucher bill (S.B. 10) authored by Senator Teel Bivins.[127] The senators vowed not to allow suspension of senate rules to bring up the bill out of order, effectively preventing it from coming to the Senate floor for debate. One senator, Gregory Luna, who was in intensive care at a San Antonio hospital, said he would leave his hospital bed and travel to Austin to prevent the bill from coming to the senate floor. Senator John Whitmire stated, "We have 11 senators who are committed to blocking school vouchers from becoming a reality in the state of Texas."[128]

The Two-Thirds Rule led Senator Bivins, chair of the Senate Education Committee who sponsored an earlier voucher bill (S.B.

1206), to conclude, "I don't see much point in voting a bill out of committee just to run into a brick wall when trying to get it up on the Senate floor."[129] "It's dead—there are 11 solid votes to block it" said African American Senator Royce West (D), a member of the committee.[130] Thirteen senators joined forces to prevent the legislation from reaching the Senate floor.[131] Hispanic Senator Gregory Luna (D) called the voucher proposal an "assault on the public schools."[132] Bivins complained his voucher plan was being stopped by a solid wall of Democratic opposition.[133]

A member of CPS was asked, "If there was enough unified Democratic opposition, given a slight Republican majority in the Senate, could they effectively block legislation?" The response was "Yes, block it from coming up." A senior staff member familiar with Senate rules and procedures stated that the Two-Thirds Rule "essentially allows any eleven senators to block any legislation. What it really takes is a two-thirds majority to pass anything. And so, while you might have a Republican majority, you know, 17 ain't two-thirds [21 would be]." As table 3.1 shows, the number of Republicans in the Senate has steadily increased, to 19 members in 2003. As a result, if all 19 Republican senators voted in support of vouchers (which is by no means a certainty, given the lack of enthusiasm for vouchers in many suburban and rural areas), only two Democrats would need to support vouchers to overcome the Two-Thirds Rule.

Several respondents commented in interviews that the Two-Thirds Rule has made a critical difference in blocking voucher legislation. One remarked, "it was totally stopped in the Senate by the vote of Democrats who just didn't let it go any further." Despite repeated attempts by Republican leaders to amend and restrict the scope of voucher proposals (such as creating a limited voucher plan targeted at children from low-income families, similar to those operating in Cleveland or Milwaukee), Democratic Senator Gregory Luna said, "Senator Bivins has reformed his measure in some ways. But most of us are still opposed to vouchers on principle."[134] A member of the anti-voucher coalition said the anti-voucher voting block is very solid and believes they will block voucher plans in the Senate again next session. He believes voucher opponents might even do better next session with the work of the Coalition for Public Schools—the broad-based anti-voucher coalition.

A senior Senate staff member stated his belief that "we probably would have had a voucher plan without the Two-Thirds Rule." A member of one of the teacher groups explicitly credited the Two-Thirds Rule with helping defeat voucher legislation in past sessions.

It was clear from interviews with several key lobbyists, legislative aides, committee staffers, and legislators that voucher opponents manipulated the procedural rules of the legislature to achieve their objectives. A member of one of the teacher organizations said, "The rules [between chambers] do vary because that's one of the things we rely on, that was our initial approach with this is that if we can just get 11 votes, we can keep this from coming to a vote [in the Senate]. That's all we had to do." The Two-Thirds Rule has even had an impact on the order with which school choice legislation is debated in the House and Senate. During the 1999 legislative session, the House Public Education Committee announced it would not even consider voucher bills until the full Senate passed a voucher bill, since there was little to be gained from the House debating such a controversial issue.[135] Pro-voucher Republicans in the House were unwilling to expose themselves to a political backlash by publicly supporting controversial legislation that had little chance of passing legislative muster in the Senate.

Although the Two-Thirds Rule is a procedural rule and is not written into the state constitution, it is one of a series of rules agreed to by members of the Senate at the beginning of every session. When asked if the rule could be changed or eliminated to increase the chances of passage of voucher legislation, particularly given the current Republican majority in the Senate, a senior Republican senator said that it was possible to change or eliminate the rule, "but every member of the Senate that I know now or that I have known in my nine years there has been in favor of keeping the blocker bill tradition." A senior Senate staff member agreed. "Yeah, the rules could change," but there exists a great deal of institutional respect for Senate rules and procedures, in addition to the fact that the rules streamline the legislative process, making the cumbersome system workable.[136] Such political traditions are highly resistant to change.

Several interviewees expressed the belief that the Two-Thirds Rule helped create a less fractious atmosphere in the Senate. A spokesperson for CPS observed differences in how the House and Senate manage conflict. She stated "the Senate is more gentlemanly and they are more interested in not having big fights, big battles, big contentious votes." A member of a group supporting vouchers agreed, stating:

> The Senate is governed by consensus. It is completely different from the other chamber in that arguments over policy generally don't take place on the floor of the Senate. They tend to take place behind the scenes. And then, by the time something makes it to the floor of

> the Senate, the way Governor Bullock[137] has liked to work it is you get enough votes to pass your bill before we even bring it onto the floor. They get a lot of things done in the Senate as a result of doing it that way. A lot of the debate takes place behind the scenes.

This tradition of the upper chamber, the Senate, governing by consensus is common among state legislatures, leading them to more readily adopt middle-of-the-road reforms with respect to school choice, such as magnet and charter schools, rather than more radical forms of choice such as vouchers.

As in many state legislatures, the institutional configuration and procedural rules of the U.S. Congress have also blocked efforts to expand school choice. For example, President Bush has repeatedly urged Congress to pass legislation providing a federal income tax credit of up to $2,500 to allow parents to send their child to a private school or another public school, as well as to offset the costs of private tutoring, books, and computers.[138] However, while Republicans control both the House and Senate, getting any type of voucher bill or tuition tax credit through the Senate (such as a voucher plan for the District of Columbia school system) will require rallying much more support than currently exists. Several senators have threatened to employ the filibuster should any such legislative proposal come to the floor for debate. Under Senate rules, 60 votes are needed to stop a filibuster. "While a handful of Democrats have supported voucher experiments before, opponents said there have never been enough to meet the 60-vote threshold, especially with a few moderate Republicans opposed."[139]

Crafting Legislative Strategy: Packaging Choice

When voucher proponents tried to get a statewide voucher plan passed in Ohio, Governor George Voinovich attempted to circumvent the traditional legislative process by incorporating a voucher plan into the state budget. According to Representative Ronald Gerberry (D-Youngstown), "The only reason this is in the budget is because George Voinovich can't pass it on its own merits."[140] Gerberry continued, "The only way to get vouchers in this state is by sneaking it into the budget."[141] Such tactics are not uncommon. State policymakers of all stripes and persuasions routinely attempt to insert controversial legislative proposals into other bills—be it the annual budget, omnibus spending bills, or as riders to other bills—all part of the logrolling process. In their study of special education policies in

Texas, Virginia Baxt and Liane Brouillette found that omnibus bills were an effective strategy to introduce and pass controversial legislation with little notice.[142]

It is also easier to pass legislation that contains something for everyone than to pass narrow, single-issue proposals, particularly on contentious issues such as vouchers. In an interview, a member of the Texas House stated, "The way we [voucher supporters] have gotten things up for a vote is to do it as an amendment when the entire education code is up for a vote. When the education code comes to the floor as being sunsetted or if we are doing a huge bill, that has a really big title on it, that's all inclusive, then you can do pretty much any kind of amendment you want to."

This procedural tactic has been used by voucher supporters to circumvent the blocking strategies of legislators opposed to vouchers. In the 1999 and 2001 Texas legislative sessions, voucher proposals were attached as riders to several bills; however, the provisions were removed from the bills before passage. Late in the 1999 legislative session, Lt. Governor Rick Perry said he would pass voucher legislation at all costs, including forcing S.B. 10 to the Senate floor "in some unique fashion" if blocked by the Two-Thirds Rule.[143] Perry was quoted as saying, "there are lots of ways to pass legislation," including attaching voucher amendments to other bills.[144] With one week left in the 1999 legislative session, Representative Ron Wilson, a pro-voucher Democrat, attempted to attach a voucher amendment to a telecommunications bill, but the proposed amendment was rejected by the House.[145]

In Wisconsin, Polly Williams, an African American Democratic assemblywoman from Milwaukee, was able to use the institutional processes and rules of the Wisconsin legislature to circumvent the traditional legislative route through which most bills flow. In an effort to avoid a potentially disastrous battle on the floor of the Wisconsin legislature, Williams and her supporters, including conservative Democrats, Republicans, and other African American members of the state assembly, had the controversial Milwaukee voucher plan "approved as part of the biennial budget process, which meant that the bill avoided the floor attention that might have made its passage more difficult."[146] The success of Williams's legislative strategy appears to be an exception to the rule, since the majority of such efforts, to date, have failed. However, given the even-greater difficulty of shepherding voucher legislation through traditional legislative channels, such efforts are likely to become more common in the future.

SCHOOL CHOICE AND ELECTIONS

Electoral cycles also affect school choice initiatives. In 1998, Michigan Governor John Engler "did not want the school choice initiative on the ballot in November" because he did not "wish to campaign on the issue during his reelection effort and risk stirring up the unions."[147] When Republicans took control of the Wisconsin state legislature in 1995, they expanded the Milwaukee voucher plan initiated five years earlier.[148] In Arizona, the election of a Republican-controlled senate majority in 1994, coupled with strong gubernatorial support, led many to believe that the state would adopt a voucher plan similar to the one operating in Milwaukee. Voucher legislation passed the state House of Representatives three consecutive years, beginning in 1992, while each time being blocked by a recalcitrant Senate controlled by Democrats.[149] Perceived deadlock in the state senate prompted key legislative leaders, including the respective chairs of the Senate and House education committees, to introduce a compromise charter school bill.[150] The bill passed "swiftly, with broad bipartisan support and almost no debate."[151] In their analysis of school choice politics in Arizona, Frederick Hess and Robert Maranto conclude that, "The dominant political position enjoyed by Republicans allowed them to use the threat of voucher legislation to squeeze union and Democratic opponents of school choice into accepting compromise charter school legislation."[152] The impact of the charter school bill has been dramatic, with Arizona leading the nation in the number of charter schools in operation.

To illustrate the impact of elections on school choice, we examine electoral trends in Texas, Pennsylvania, and New York. These states share several commonalities: (1) each has a large population and plays an important role in national politics; (2) each was led by a popular Republican governor within the past decade; and (3) each legislature initiated charter school legislation and two of the three states (Texas and Pennsylvania) have experienced heated, intensely partisan battles over vouchers.

Texas

As table 3.1 shows, the Texas legislature has become steadily more Republican in the past two decades. In 1982, Democrats controlled 76 percent of the seats in the state house; by 2003, Republicans held a significant majority for the first time in over 100 years. In the state Senate, Democratic control has dwindled from 84 percent (1982) to only 39 percent in 2003. As table 3.1 shows, since 1982, the number of Republicans in the House has more than doubled, from 36 to 88, while the number of Republicans in the Senate has increased nearly four-fold. Republicans hold majorities in both chambers for the first time since the end of *Reconstruction*.[153] At no time in Texas's post-Reconstruction history has the Republican Party had such numerical strength in the legislature.

In interviews with state policy elites, all expressed the view that the changing composition of the legislature is helping the voucher movement. A Republican state representative who strongly supports vouchers said "it [the increase in the number of Republicans in the state legislature] has made a huge difference" for the prospects of passing vouchers. A senior official for one of the teachers' groups stated, "This whole thing [vouchers] has gotten, if you go back to '91, '93, and '95 and every year the vote gets a little closer." A senior staff member who served on one of the education committees stated,

> I think what is happening is clearly there is, I think, we're seeing, independent or irrespective of what your perspective is on this, I think we are seeing a shift toward increasingly embracing vouchers. I think that shift is probably predicated to some extent or tied to the same shift we are seeing as Texas makes the full transition from a state that had heretofore been dominated by the Democratic Party to a state that now is essentially dominated by the Republican Party. Republicans are clearly much more amenable to the concept [vouchers] than generally speaking Democrats have been.

This statement is supported by voting records of the last four legislative sessions. For example, in three record votes on private school vouchers during the 1997 legislative session, only four House Republicans voted against vouchers, and two of these reportedly did so only because a pro-voucher group contributed large sums of money to their opponents in the Republican Party primary. With a few notable exceptions, the voucher movement in the state legislature is sharply divided along party lines. Nearly all minority lawmakers in the legislature oppose voucher plans, although with only 12 Democrats in the Senate, a few defections could be enough to get voucher legislation through the Senate. As a member of a teachers' group said in an interview, "Most of the African American [legislators] in the House are still with the coalition [CPS] on this issue." The vocal opposition of nearly all the Hispanic members of the legislature is particularly significant in a state with the second-highest percentage of Hispanic legislators.[154] Nearly 20 percent of the legislature is Hispanic, nearly 10 percent African American.

In 1994, Republican challenger George W. Bush defeated popular incumbent Democratic Governor Ann Richards. Bush's election was significant insofar as it inaugurated only the second Republican gubernatorial administration in Texas since the end of Reconstruction.[155] Voucher plans had been blocked in the past by state Democratic leaders, but Bush's victory over Democratic incumbent Ann Richards, a voucher opponent, led to a revival of such plans.[156] Democrat State Representative Ron Wilson, who co-sponsored a voucher bill during the 1993 legislative session, called Bush's election "a great sign" for vouchers.[157] Wilson said "what we have to do now is work in the House and Senate to get a consensus on a pilot program. His [Bush's] election will help us get that done."[158] State Senator Gonzalo Barrientos, however, noted that "the governor doesn't run the state" and his initiatives must pass through a Texas legislature under Democratic control (in 1994).[159] While vouchers did not become a reality under Governor Bush, a strong charter school law was passed the year after Bush became governor in 1994. Changes in the systemic governing coalition in Texas politics, under unified Republican control for the first time since the end of Reconstruction, greatly increases the chances that a pilot voucher plan will pass legislative muster in 2003, although this outcome is by no means certain.

Pennsylvania

The Pennsylvania legislature has also come under greater Republican influence over the past two decades, although the longitudinal electoral data suggest greater interparty electoral stability than in Texas. As table 3.2 shows, Republicans held a slim majority in the state house in 1982, and have increased their majority slightly since then. The state senate has been under Republican control since 1982, although Republican dominance increased only slightly (by three seats) during that period. Historically, even when one party has controlled both houses of the Pennsylvania legislature, "its majority has usually been narrow."[160] As a result, "the party in power has always had to accommodate the minority, whether by working with the minority leadership or by dealing with selected individual members of the minority."[161]

School choice initiatives were given a significant boost in 1994 when Republican Tom Ridge, an ardent voucher advocate, captured

Table 3.2 Pennsylvania election results, 1982–2003

	1982	1988	1994	2003
House				
Democrats	100	99	102	94
Republicans	102	104	101	109
Senate				
Democrats	23	23	24	21
Republicans	26	27	26	29

Sources: John J. Kennedy, *The Contemporary Pennsylvania Legislature* (Lanham, MD: University Press of America, 1999), p. 7; www.ncsl.org/statevote98/tottrn.htm; www.house.state.pa.us; www.pasen.gov; www.ncsl.org/programs/legman/elect/statevote2002.htm.

the governor's mansion. During his eight years as governor, Ridge tried repeatedly to push various voucher plans through the state legislature, with little success. Within the state legislature, "school choice has never been an issue that easily divides along party lines."[162] If it were, then school vouchers would have passed in 1995, when Governor Ridge, an ardent voucher proponent, first proposed it.[163] The governor was successful in shepherding a reasonably strong charter school bill through the legislature in 1997. The politics of negotiation and compromise produced by slim governing majorities have played a major role in shaping school choice policies in Pennsylvania, with policymakers more likely to adopt less radical reforms such as charter schools, and, in the case of Philadelphia, state takeover and reconstitution of failing school systems.

New York

New York has not experienced the same degree of conflict over school choice as Texas or Pennsylvania, due in part to remarkable stability in state electoral politics. As table 3.3 illustrates, Democratic and Republican strength in the state Assembly and Senate has been nearly unchanged since 1982, with Democrats dominating the Assembly and Republicans in firm control of the Senate. Despite the *Zelman* decision, the Democratic stranglehold on the state Assembly makes it highly unlikely that a voucher plan will be adopted in New York in the foreseeable future. In fact, Democratic control of the state Assembly has increased since 1982, although only by about 4 percent (six seats). As in Texas and Pennsylvania, the 1994 elections

Table 3.3 New York election results, 1982–2003

	1982	1988	1994	2003
Assembly				
Democrats	97	92	94	103
Republicans	52	58	56	47
Senate				
Democrats	26	27	25	25
Republicans	35	34	36	37

Sources: Edward Schneier and John B. Murtaugh, *New York Politics* (Armonk, NY: M. E. Sharpe, 2001), p. 99; www.ncsl.org/programs/legman/elect/statevote2002.htm.

Table 3.4 Political party control of state legislatures, 1981–2003

	1981	1985	1990	1995	2003
Democrats	28	27	29	18	16
Republicans	15	11	9	19	21
Split	6	11	11	12	12

Note: Totals do not equal 50 because Nebraska has a unicameral, nonpartisan legislature.

Sources: *Statistical Abstract of the United States, 2000, 2001* (Washington, D.C.: U.S. Bureau of the Census, 2000, 2001); www.ncsl.org/programs/legman/elect/statevote2002.htm.

swept a Republican into the governor's mansion in New York, with George Pataki defeating incumbent Governor Mario Cuomo, who had held the office the previous 12 years. School choice initiatives had been on the backburner of Albany's legislative agenda, but Pataki's election renewed hope among choice activists that significant progress could be made. Four years later, largely due to Pataki's insistence, the legislature passed a strong charter school law.[164] Vouchers have never been seriously considered by Governor Pataki because the governor is not a vociferous advocate of vouchers and the chances of even a small pilot voucher plan passing through the legislature are virtually nil.

As the data in the tables above indicate, the 1990s witnessed a period of growth in Republican power in the state legislatures, which brought increased opportunities to expand school choice initiatives. As shown in table 3.4, Republicans held majority control over both chambers of the state legislature in only nine states (18 percent of the total) in 1990, but only five years later, they took control over

Table 3.5 Governors by party affiliation, 1981–2002

	1981	1985	1990	1995	2002
Democrats	31	34	29	19	24
Republicans	19	16	21	30	26
Independent	—	—	—	1	—

Sources: *Statistical Abstract of the United States, 2000, 2001* (Washington, D.C.: U.S. Bureau of the Census, 2000, 2001); www.nga.org.

19 state legislatures (39 percent). In 2003, Republicans held majorities in the legislatures of 21 states. The growth of Republican power in state and national politics in the past twenty years has greatly benefited the school choice movement. In Ohio, the shift in the balance of power from Democrats to Republicans has had a significant impact on school choice initiatives. In 1988, Democrats held a 59-40 majority in the House of Representatives, while Republicans controlled the Senate 19-14. However, in 1994, Republicans captured control of the House (56-43), and extended their control in the Senate (20-13).[165] It is no coincidence that Ohio's school choice program, the Cleveland voucher plan, passed the state legislature one short year after Republicans took control of the statehouse.

The decade of the 1990s also brought about a dramatic increase in Republican gubernatorial control. As table 3.5 illustrates, between 1990 and 1995, the number of Republican governors increased 18 percent, from a minority of 21 (1990) to a majority of 30 (1995). Currently, Republicans occupy a slim majority of governorships nationwide. As with changes in the composition of state legislatures, changes in gubernatorial control affect the opportunities for school choice advocates to advance their cause. Republican governors tend to be more receptive to various forms of school choice than Democrats, which sometimes translates into expanded choice initiatives. As Bryan Hassel found in his analysis of the effects of gubernatorial leadership in school choice, Republican governors are only slightly more likely than Democratic governors to pass charter school legislation, although they are significantly more likely to pass strong rather than weak charter laws.[166]

CONCLUSION

Since Republicans overwhelmingly, although not uniformly, support school vouchers, changes in the composition of political actors (the systemic governing coalition) during the past two decades have significantly increased the prospects for expanding school choice, particularly radical forms of choice such as vouchers. The strengthening of the neo-conservative position in U.S. politics "has made some privatization possible and has put vouchers within reach."[167] However, despite these dramatic electoral trends, significant institutional constraints exist that substantially slow the pace with which choice reforms are adopted and implemented. For example, despite having the Texas Senate under Republican control for the first time in over a century, voucher plans failed to make it through the state senate during the 1997 legislative session. Pressure by a pro-voucher Republican governor—one of only three Republican governors of Texas in over a century and who personally lobbied legislators on behalf of a pilot voucher program targeted at disadvantaged children—was insufficient to push voucher bills through the legislature.

As this chapter has shown, the institutional context of state policymaking exerts a significant impact on policy adoption and change, particularly on hot button issues such as school choice. Although the institutional structure "is not an independent determinant of policy outcomes," it does serve as "a conditioner of the political processes by which those policy outcomes are achieved."[168] Further, this institutional context, with its built-in resistance to radical departures from existing policies, indicates a bias against the most controversial forms of school choice such as vouchers, increasing the likelihood of passing charter school legislation, often at the expense of vouchers. The Texas legislature in particular is structured in such a manner that policy initiatives that do not have widespread support cannot pass through the legislature. This structure facilitates passage of noncontroversial, middle-of-the-road policy initiatives while blocking more radical reforms—in essence, incremental policymaking—a common outcome in a federal system of governance with separated powers and numerous checks and balances.

Thus, far from being a neutral arbiter or "simply a neutral converter of societal demands,"[169] the institutional structure of the state plays a key role in shaping the political dynamics of school choice. Although this institutional bias does not preclude the adoption and implementation of radical policy initiatives, the institutional context clearly makes passage of such plans difficult, since the process is designed to forge consensus and discourage radical policy initiatives. An old Texas proverb sums up this bias well: "No one's life, liberty, or property are safe while the Texas legislature is in session."

CHAPTER 4

INTEREST GROUP DYNAMICS: THE POWER OF COALITIONS

While institutions play a key in shaping and influencing the outcome of the policy process, many scholars assert that, ultimately, it is people—policymakers, activists, and interest groups—who determine the outcome of policy initiatives. Moving away from outmoded conceptions of iron triangles, neopluralist scholars examine the complex workings of policy subsystems—particularly the formation, alignment, and realignment of coalitions within and between those subsystems. Paul Sabatier observes that "one of the conclusions emerging from the policy literature is that understanding the policy process requires looking at an intergovernmental policy community or subsystem" as the basic unit of study.[1] A policy subsystem is defined as "those actors from a variety of public and private organizations who are actively concerned with a policy problem or issue," including "actors at various levels of government, as well as journalists, researchers, and policy analysts."[2] Research on policy subsystems has become the dominant paradigm of interest group scholars, and a burgeoning literature has developed around it. An anthology of classic works in public policy by Daniel McCool devotes an entire section to policy subsystems.[3]

Varying degrees of consensus, conflict, cooperation, and competition exist within these policy subsystems. Often, coalitions form within policy subsystems—each seeking to obtain desired benefits through changes in programs and policies. In an attempt to explicate the political dynamics of this process, Hank Jenkins-Smith and Paul Sabatier developed a model of interest group interaction that they call the advocacy coalition framework. Within policy subsystems, interest groups—and the policies and programs they seek to promote—have well established belief systems that are relatively stable over time, much like models of political culture discussed in chapter 2. Sabatier and Jenkins-Smith argue that policy change and learning can best be understood as the product of competition among interest groups within the constraints of a policy subsystem.[4] They argue that

> policy change over time is a function of three sets of processes. The first concerns the interaction of competing *advocacy coalitions* within a policy subsystem. An advocacy coalition consists of actors from a variety of public and private institutions at all levels of government who share a set of basic beliefs (policy goals plus causal and other perceptions) and who seek to manipulate the rules, budget, and personnel of governmental institutions in order to achieve these goals over time. The second set of processes concerns *changes external to the subsystem* in socioeconomic conditions, system-wide governing coalitions, and output from other subsystems that provide opportunities and obstacles to the competing coalitions. The third set involves the effects of *stable system parameters*—such as social structure and constitutional rules—on the constraints and resources of the various subsystem actors.[5]

According to this model, networks of policy actors learn how best to play the political game to achieve their policy objectives. Sabatier and Jenkins-Smith refer to this process as policy learning. Policy learning is different from organizational learning (discussed in chapter 5) wherein policymakers, operating in organizations, utilize systemic learning processes to improve policies. Policy learning, as used by Sabatier and Jenkins-Smith, is really political learning—an altogether different process. Sabatier argues that "on the basis of perceptions of the adequacy of governmental decisions and the resultant impacts as well as new information arising from research processes and external dynamics, each advocacy coalition may revise its beliefs or alter its strategy."[6] Thus, learning is conceptualized as the product of each actor's (or interest group's) success at achieving its intended policy objectives.

This approach has several advantages over the rational actor approach common in studies of organizational learning (see chapter 5). Joseph Stewart argues that "in looking at change in public policy, particularly over an extended period of time, the advocacy coalition approach brings some important elements more explicitly into the analysis than does the rational actions approach. In particular, the role of new information, ideas, or assumptions can be considered in ways other than just as rational actions within organizations."[7]

Another advantage of this approach to understanding policy change is that advocacy coalition models incorporate a change component into the theory, freeing it from the static, status quo elements of previous theories. By focusing on changes external to the policy subsystem, such as changes in socioeconomic conditions, governing coalitions, and outputs from other subsystems, advocacy coalition models demonstrate how changes in the external environment impact the policymaking process.[8] Support for this model comes from studies of the dynamics of the legislative process. According to Kevin Hula, "Organized interests fight their major battles today largely in coalitions."[9] Sabatier found that policy change occurs via the interaction of competing advocacy coalitions.[10] Frank Baumgartner and Bryan Jones believe that the coalition-building component of the theory is essential for policy change and innovation.[11]

ADVOCACY COALITIONS AND THE POLITICS OF EDUCATIONAL CHANGE

A number of studies of educational change support advocacy coalition theory. In their review of major multiple-state case studies in education, Martin Burlingame and Terry Geske assert that "the politics of education at the state level is still a politics of interest groups."[12] In her studies of policy change in Canadian education, Hanne Mawhinney found that "the Ontario educational policy community is tightly knit with well defined sets of assumptions and norms."[13] This suggests that policy communities have a significant impact on the nature and direction of policy change. Many such communities exist within the school choice movement. For example, the Home School Legal Defense Association and the National Home Education Network have been instrumental in the drive to legalize home schooling throughout the United States. The success of this movement attests to the power of organized interests in shaping education policy at the state level.

Advocacy coalitions have emerged as powerful players in the educational policy arena. In his analysis of changes in state educational policymaking over 20 years in Minnesota, Tim Mazzoni found that advocacy coalitions were a driving force behind the educational reform movement.[14] Mazzoni argues that Minnesota's state school policy subsystem can be characterized as an advocacy coalition of innovative reformers, which contributed significantly to Minnesota's adoption of the nation's first charter school law in 1991. He observed that "linking together government, business, education, foundation, parent, and civic actors—and led by elected officials—this coalition became a potent force in setting forth a restructuring agenda and in influencing the policy system to adopt public school choice as the central element in that agenda."[15] According to Mazzoni, Minnesota's advocacy coalitions "have repeatedly squared off during the past decade over issues of school reform, with their struggle appearing to have been spawned by a fundamental cleavage over core beliefs, by stable structural features of the institutional setting, and by the impact of multiple changes in a turbulent external environment."[16] Mazzoni concludes that Sabatier's advocacy coalition model is a useful approach to understanding policy change in education and that it "appears to fit significant developments within Minnesota's education policy system."[17]

The findings of Mazzoni's research are consistent with Robert Feir's analysis of education policymaking in Pennsylvania.[18] Feir found that a coalition of business leaders, media, governors, and chief state school officers were actively engaged in educational reform, while traditional education interest groups played minor roles in the reforms of the 1980s. He notes that "the expansion of the conflict over education reform to include business, political, and media leaders, coupled with the substantial neutralization of education interest groups, provided opportunities for new actors to set the agenda."[19] The Pennsylvania State Education Association (PSEA), the state's largest teacher's union and traditionally the most influential lobbying organization in the state, lost considerable power in the 1980s and 1990s when it opposed two popular Republican governors—Richard Thornburgh and Tom Ridge—leading to "a long period of isolation from the administration's policy discussions."[20]

Changes in the composition of actors in policy subsystems, coupled with changes in the external environment, often play a determinant role in opening up windows for policy entrepreneurs to initiate policy change.[21] Michael Mintrom found that policy entrepreneurs such as Joe Nathan in Minnesota, Polly Williams in Wisconsin, and Paul DeWeese in Michigan "raise significantly the probability of legislative consideration and approval of school choice as a policy innovation."[22] In 1998, a tax credit initiative in Colorado and charter school legislation in California "both began as ballot initiatives and each was the brainchild of a businessman."[23] As Hubert Morken and Jo Renee Formicola observe, "What is new in school choice is the arrival of entrepreneurs—activists who are independent, freewheeling, sensitive to marketing issues, and able to move with lightening speed and chutzpah."[24] In fact, much policy change in education, particularly educational reform, comes from outside the traditional educational policymaking subsystem.[25]

INTEREST GROUP DYNAMICS AND SCHOOL CHOICE

To evaluate the saliency of interest group activism and the relevance of advocacy coalition models of policy change on school choice, I examine charter schools and vouchers separately, since in many ways the politics surrounding these forms of school choice are quite distinct. I begin the analysis by examining the political debate surrounding various attempts to pass voucher legislation.

Vouchers, Interest Groups, and Advocacy Coalitions

Of recent state-level educational reforms, none generate as much conflict or hortatory political rhetoric as proposals for educational vouchers. In no arena are the divisions between supporters and opponents of school choice more clearly drawn. For this reason, vouchers are the ideal issue from which to examine interest group activism and competition among advocacy coalitions.

In Texas politics, various interest groups coalesced into well-defined advocacy coalitions over vouchers. Members of the pro-voucher coalition (see Appendix A) advocated vouchers as a method to "equalize educational opportunity by giving all families the options now enjoyed only by those wealthy enough to send their children to private schools or move to areas with high-quality public schools."[26] Speaking in support of vouchers, African American state representative Ron Wilson stated "we're talking about the people who fund the system having the option to take some of their tax dollars and apply them to their child's education at a private institution."[27] Wilson stated that "minority parents feel trapped. Until we fix what's wrong with the schools, they need to have options."[28] Wilson asserted that "parents are looking for a tool to give their children a fighting chance."[29] Republican Senator David Sibley (R) noted, "you're talking about people who will be making decisions for their own children."[30]

In an editorial, Republican state representative Kent Grusendorf, who introduced a pilot school choice plan for low-income students during the 1993 legislative session and was a member of the House Education Committee, asserted that "Texas parents are the best qualified to determine which school best fits their own child's individual needs. Where more school choices are available, parents will determine which school best meets their own child's educational needs."[31] Senator Jane Nelson (R), like Wilson and Grusendorf a supporter of a pilot voucher program for disadvantaged students, argued that "the bottom line is the (voucher) recipients would be children of low-income families...in inner-city school districts, who have no other way out of the system."[32]

Jimmy Mansour, president of Putting Children First (PCF), a pro-voucher organization, argued "parents (should be) given the ability to do everything in their power to get a high quality education in a safe environment."[33] According to public records, PCF spent $650,000 on lobbying efforts during the 1997 legislative session.[34] Cecile Richards of the Texas Freedom Network, a liberal watchdog group, criticized the voucher proposal as a "billion-dollar experiment local taxpayers can't afford."[35] She said Mansour and his political action committee raised "over $500,000 in the past year to elect extremists to the State Board [of Education] who don't even send their own children to public school."[36] A campaign finance report released by the anti-voucher Texas Freedom Network's Education Fund reported that pro-voucher organizations contributed over $5.2 million to pro-voucher candidates and PACs.[37] James Leininger alone contributed nearly $4 million to various political campaigns, including a $1.1 million loan to Rick Perry (then running for lieutenant governor) and a $950,000 loan to Carole Keeton Rylander (then running for state treasurer).[38]

Mansour claimed choice was a "unifying [issue], one that brings all folks together, black, white, brown, liberal, conservative."[39] Reflecting the broadness of this coalition, then-Lieutenant Governor Bob Bullock, a Democrat, was named honorary chair of the pro-voucher group. Bullock also served on the group's legislative advisory board, whose membership included several other Texas legislators. A bipartisan group of legislators supporting PCF includes Democratic state Representatives Henry Cuellar, Ron Wilson, and Glenn Lewis; Republican Representatives Tom Craddick (the Speaker of the House in 2003), Kent Grusendorf, and Mike Krusee; Democratic Senator Ken Armbrister; and Republican Senators Jeff Wentworth and Jane Nelson.

In an interview, a member of another major voucher advocacy group stated, "They're [Putting Children First] only goal is to see educational choice in Texas." Another major voucher advocacy group is the Texas Public Policy Foundation (headquartered in San Antonio), which an interviewee stated "was the organization responsible for making choice an issue years and years ago back when no one had even heard of it [in 1989]." One informant identified the Texas Justice Foundation and Putting Children First as "the biggest pro-voucher advocates." He also noted that "the Catholic schools have their own individual problems, and they are probably more supportive of vouchers."

Geraldine Green, an African American parent who sends her son to a private school, supports efforts to establish vouchers. She stated, "What we're saying is that our tax dollars can go toward private education, and it should be the parent's choice... Too many of our black male children are falling through the cracks. Why not let us use our tax money to give them a better chance?"[40] Green asserted "I don't think they [minority legislators] understand how important this is to parents. Either they're not listening or they're not concerned."[41] Another African American parent who also sends her son to a private school stated, "It's a really big [financial] struggle, but I refuse not to provide the education he needs and deserves. Many parents are like myself and can't afford private schools. We should be able to take our kids wherever they will be educated."[42]

Republican state representative Kent Grusendorf noted that "a privately funded pilot voucher program is available to disadvantaged students in several Texas cities."[43] This privately funded voucher program is called the CEO Foundation and, according to an informant, "was formed out of TPPF [the Texas Public Policy Foundation]" in 1992. The CEO San Antonio Foundation provides more than $500,000 in vouchers to 950 low-income students in grades one through nine in Bexar County.[44] The foundation was founded by wealthy San Antonio businessman James Leininger, who donated $1.8 million to the foundation (80 percent of its funding).[45] The foundation is "one part philanthropy and one part demonstration project to convince the Texas Legislature to create a taxpayer funded private school voucher program."[46] Since 1992, 30 private voucher programs have been established around the country with $40 million in funding from donors.[47] Each program is designed to create public support to pressure state legislatures to adopt vouchers.[48]

Vocal opposition to voucher plans arose from a variety of groups not normally in agreement with one another. Carolyn Boyle, spokesperson for the Coalition for Public Schools (CPS)[49] (see Appendix B), which helped defeat a voucher bill in 1995, argued that "Texas can't afford a $1 billion private school voucher experiment that would drain money from underfunded public schools."[50] CPS was created in 1995 for the sole purpose of opposing vouchers. A spokesperson for the organization said, "opposing vouchers was why it [CPS] was created. We are a one-issue coalition." A poll released by CPS showed 67 percent of Texans think public tax dollars should be spent only to improve public schools; 27 percent believed public tax dollars should be used to assist parents who send their children to private, parochial, or religious schools.[51] Boyle stated "we believe public money should only be spent for the support of public schools, that the budgets are very tight on every school campus in Texas."[52] The Association of Texas Professional Educators (ATPE) estimates that local public schools could lose up to $5,000 for each student using a voucher at a private school.[53]

Members of the Texas PTA stated their opposition to "using public dollars for private schools" as did the Texas Classroom Teachers Association.[54] A senior official with TEPSA said in an interview, "It seems to be very much a right-wing agenda. You also have folks that are very much pro-business, kind of this free-market competition approach that will improve our schools. You have the religious right, who are very much pushing it. And then we had groups of minority legislators (Cuellar and Wilson) that joined forces, mostly some minority Democrats had joined forces with these ultra-conservative Republicans." The official went on to say "if you look at the Christian Right things that are put out by certain folks, their mission is to eliminate the public schools. It really is." She noted that many noneducational motives are at work in the battle over school vouchers. "You have motives, there are people involved that simply want individual rights. That puts public education in a different, it gives it a different purpose than what it's had."

Members of the Texas State Teachers Association (TSTA) expressed concern that choice programs could further widen the gap between rich and poor.[55] TSTA President Richard Kouri sharply criticized voucher plans, calling them the "agenda of the extreme right and other ultra-conservative groups."[56] In an interview, an official with TEPSA noted that "good private schools cost a lot more money than a voucher would ever be worth. So, it's only going to benefit the real wealthy and the upper-middle class." Gil Gamez, director of the League of United Latin American Citizens (LULAC), said, "Hispanics are waking up to the fact that so-called reform proposals are only disguised efforts to shut out minorities from educational and economic advancement."[57] Several religious groups throughout Texas, particularly Baptists, oppose vouchers, although the Catholic Church is generally supportive of such plans. According to a member of CPS, the "Catholic Church had all kinds of bishops and people over here lobbying for vouchers because the Catholic schools need more money [due to fewer nuns teaching, and thus having to pay higher teacher salaries]."

Fearing anyone or any organization could create schools, Brad Duggan, executive director of TEPSA, questioned "whether religious zealots or political extremists could set up schools with state funds."[58] In an interview, another official with TEPSA stated "vouchers don't seem to have that degree of accountability. David Koresh could open a school in Waco. You could have anything going on." Duggan went so far as to criticize the motives of voucher proponents. He asserted that "the whole voucher program has been sold as [the] way for low-income parents to obtain the same quality of education as the governor's children. But we know private schools aren't going to open their doors to all students."[59]

As is the case in several states throughout the country, the traditional public education establishment set aside their internal differences and came together in an effective advocacy coalition to block vouchers. In testimony before the Senate Education Committee, Linda Bridges of the Texas Federation of Teachers said:

> TFT regards private school vouchers as a very serious threat to the interests of Texas school children. TFT is a member of the Texas coalition in opposition to private school vouchers which now includes every major public education organization in Texas [CPS]. Article 7, Section 1 [of the state constitution] clearly states that it shall be the duty of the Legislature of the state to establish and make suitable provisions for the support and maintenance of an efficient system of education. It is not the duty of the Catholic Church or the Baptist Church or any private academy. We understand the political appeal of choice as a concept. It is always smarter to campaign for choice than for no choice. Yet, private school voucher systems are not sweeping the nation. In every instance where voters have had the opportunity to vote in a referendum on this issue, they have rejected vouchers.

In an open letter to the legislature, Senator Mario Gallegos, Jr. (D) stated he had "received many calls and letters from constituents voicing their concerns about spending public money on private education." He voted to strike voucher language from S.B. 1 during Senate debate.[60]

Several legislators who supported charter schools remained firmly opposed to voucher plans. African American state Representative Wilhelmina Delco (D), a supporter of charter schools, staunchly opposed vouchers for use in private schools.[61] Representative Sylvester Turner (D), an African American legislator from Houston, stated "when people want something, they don't mind using minorities and the poor to open up the door for programs they want. If you want it, use your own kids to justify it."[62] Turner asserted "if you are poor, if you are a child, if you have been abandoned—nobody speaks for you and you are left out of the process."[63] Voucher plans were criticized by lawmakers as returning segregation to public schools, making them separate and unequal.[64] Senator Royce West (D), an African American, argued that "this [voucher bill] will create flight from the inner city. It is very clear to me . . . that if you have a low-performing school and you give that school the necessary resources, you will see improvement in student performance."[65]

An overwhelming majority of the editorials in state newspapers from 1993–2002 reveals outspoken opposition to various voucher plans. These same newspapers had earlier been filled with endorsements of charter schools. In an editorial in a major state newspaper, it was argued that "not every child can take advantage of vouchers or transfers within public schools because there are only so many schools and so many desks."[66] Some expressed the fear that vouchers "could work against social mobility for all children because money would be drained from many districts."[67] Voucher proponents ignore or "sidestep the larger social consequences of such a system."[68] One editorial noted that, "Even with a voucher system, the poorest, least informed parents likely will keep their children in a financially strapped and educationally feeble public school system."[69] It was argued that the "solution is not to divert dollars from public education to private schools, but to continue to champion more money for public education [and more accountability] and utilize existing education reforms to improve the system."[70] One writer asked, "Why subsidize a policy, under the guise of improvement through competition, that could end up subsidizing middle-class flight from the public school system and reduce the role of education in attempting to establish a level playing field for all Texans?"[71]

Some activists expressed opposition to even a small-scale pilot voucher program. Phil Strickland of the Texas Baptist Christian Life Commission stated that "some of the legislation . . . may indeed just be a baby in terms of the impact on public education. But it's a baby that's going to grow into a 500-pound gorilla."[72] In testimony before the Senate Education Committee, Richard Kouri of TSTA stated, "We think that the particular proposal (vouchers targeted at minority students) in our opinion is the nose of the camel coming in the tent, that at some point someone is going to try and push the rest of the camel through." Voucher opponents feared that once created, a voucher plan would be rapidly expanded throughout the state.

During the 1997 legislative session, private school voucher plans, on which conservative Republican spent more than $1 million on legislative campaigns and a huge lobbying effort, never made it out of either the House or Senate.[73] According to Richards, "Despite their effort to portray the voucher campaign as a way to help poor kids, the money and history associated with the voucher supporters made that hard to believe. Many of the same folks supporting vouchers had fought against earlier education initiatives for low-income children including decreasing class sizes, increasing early childhood education and equalizing school funding."[74] Michael Williams, president of the Galilee Group, a Fort Worth think tank focusing on education and equity issues, stated "both [sides of the debate] are looking at vouchers from a sense of power. Mom and dad are trying to gain control over their children's education and increase academic achievement. Minority leaders are looking to maintain institutional power—school boards, teachers associations and public schools—as the primary education provider."[75]

In Texas, as well as nationally, the most fascinating aspect of studying advocacy coalitions and interest group battles over school choice revolves around the complex issue of power. Generally, each side believes the other is more powerful and exerts more influence. Both coalitions engage in intense lobbying campaigns (at both the state and grassroots levels), contribute to political campaigns, and make their presence felt at the ballot box. Both sides claim to be vastly outspent by their opponents and insist their opponents are better organized. Each side claims to be the underdog and both claim to represent the will of the people. For example, Senator Bivins, a voucher proponent, said "they [education groups opposed to vouchers] have a lot of influence among members in the Legislature, and I'm sure that legislators are going to listen to them."[76] Representative Henry Cuellar, one of the few minority Democrats who support vouchers, said, "Without a doubt, the teacher groups and public education groups do have an impact."[77]

In 2000, the National Education Association voted to increase union dues by five dollars per member to fight vouchers in the states.[78] Reflecting in large part a state-by-state commitment to stopping vouchers, the NEA's "contributions to state ballot campaigns jumped from $1.7 million in the 1994–1996 election cycle to $6.2 million in the 1998–2000 cycle," according to a NEA briefing paper.[79] Tim Mazzoni agreed, noting that by 1993, education interest groups and their allies "had already put together blocking coalitions of formidable strength, coalitions which could draw upon the resources of state organizations, particularly the powerful teacher unions, and—if pressed—upon the financial and political resources of their national associations and networks."[80]

However, a member of CPS believed the pro-voucher coalition had much more money, power, and organizational capability. The spokesperson believes the pro-voucher groups have a lot of influence in the legislature. He stated "they [pro-voucher groups] are much more tricky and strategic than we are because they have enough money to bring in all these consultants and just sit around all day and plan their strategy. Plus, it's a nationwide strategy that is just being implemented in Texas." A senior staff member concurred, noting that the pro-voucher groups "are putting a lot of money into it." The Texas Classroom Teachers Association (TCTA) noted that Putting Children First reportedly spent $675,000 in its attempt to pass a voucher bill during the 1997 legislative session. A member of another anti-voucher group agreed, stating that the pro-voucher groups "are extremely well organized. They are very well funded and they are very focused. One thing that they have done, even going back three, four sessions ago is that they have maintained their effort. We know the opposition, they don't take off just because nobody's in Austin. They are very focused."

In an interview, a member of TEPSA said, "Putting Children First is supported by what's called the A+ Pac and there's a tremendous amount of out-of-state money" contributed by James Leininger, Jimmy Mansour, and John Walton (the head of Wal-Mart who has donated large sums of money toward voucher efforts in Washington, Colorado, and Texas). The TEPSA member interviewed stated that although she was not sure of the actual amount, "it's money that the anti-voucher people like us don't have." She said, "it's a tremendous amount of money and a tremendous influence. A+ Pac basically supported all the people on the State Board of Education you hear about being part of the religious right." According to public records, Leininger contributed $281,000 to the A+ Pac for Parental School Choice. This amounted to 53 percent of the PAC's funding.[81]

In an interview, a state senator identified the Texas Public Policy Foundation (TPPF) as the main voucher advocacy group. He does not think they have much credibility "but they certainly keep the rhetoric up." TPPF was founded in 1989 by Leininger. The foundation has issued several reports providing empirical evidence for school choice and is guided by the core principles of limited government, free and competitive markets, and individual responsibility. Leininger has given the foundation more than $980,000—about one-fifth of its total operating budget.[82] According to public records, Leininger contributed more than $700,000 between 1994 and 1996 to candidates and political committees dedicated to passing a voucher program.[83]

A senior committee staffer believes both voucher supporters and opponents exert influence over the legislature but in different ways. He said, "They have some influence [pro-voucher groups]. They have got a lot of money. The pro-voucher people have more financial backing. The school people have their organizations and they are powerful because they represent school boards and school administrators and teachers and they have a big voice because they can contact their representatives." A member of one of the teacher groups acknowledged the power of numbers when she stated, "I think that as an association, by ourselves, I won't say that it's [interest group pressure on the legislature] a lot, but I think in conjunction, we have got such a large membership so we can call those people out and say, you know, you need to call your representative or your senator. And then if they don't vote, then we can remind people of that when the next election comes around."

A member of an anti-voucher group stated that one organization, "TSTA, makes large contributions to political candidates, that weighs in." He went on to assert:

> As a block, it's [the anti-voucher coalition] pretty powerful. I dare say that were it not for the education lobbies, we probably would have had it [vouchers] in '97, if not '95. Let's just say we sat back and let it go and were neutral on the whole issue. I think we would have already had it. No question. Because when you consider all those teacher groups, and then you got the school board association, and they've got deep pockets over there, I think we would have had them.

One state representative asserted that the anti-voucher coalition has enormous power "because they have the political network. They have a proven get-out-the-vote apparatus and they have money." They do it, he stated,

> with money and they do it with their influence over the votes of people in the education industry. In an awful lot of districts, education is the number one industry in the district, especially in the rural areas of Texas. Traditionally, the voucher proponents, the only way for them to get votes is through ideology. And even that is kind of . . . Let's face it. For the most part, Republicans represent people who live in the suburbs. People who live in the suburbs have some of the better schools in the state. Republican constituents are really not clamoring

> for vouchers at all because they have a decent school system. When you get into the more affluent Republican precincts or districts, they are actually against vouchers. One is the fear of regulation of private schools. Another is there is some racism in those areas where they don't want black and brown kids going to their private school.

Although coalitions supporting school choice are usually portrayed as being highly unified, in reality significant differences exist, even among groups supporting vouchers. One state representative asserted, "The pro-voucher groups have to decide what they are going to do. Everybody has a different plan and it's hard to get everybody on board on that. When you are just against everything that brings any kind of market system into education, it's a lot easier." Some pro-voucher groups are strongly ideological and have an "all or nothing" attitude. Some groups favor vouchers for everyone, regardless of income, and favor reimbursing parents who already send their children to private or parochial school—a wholly unrealistic policy, given the enormous expenditure such plans would require. Other groups and individuals are more pragmatic, willing to make political tradeoffs to get a small-scale voucher plan targeted at poor families and their children through the legislature. The wide variety of choice plans that have been proposed in state legislatures throughout the country indicates the diverse interests and objectives of voucher supporters.

Similarly, differences exist in the stances taken by traditional education groups (administrators, teachers, and school boards) on school choice, although vouchers tends to be a useful polarizing issue upon which the diverse groups that constitute the educational establishment can agree. A member of CPS acknowledged the difficulty of keeping unity among a coalition of 25 organizations with diverse interests. He stated "you have a coalition of 25 diverse organizations [school boards, superintendents, teachers, etc.] and lots of those groups don't trust one another very much. And there also had been just lots of water under the bridge between the four teacher groups; you know, the four teacher groups all compete." However, he said the coalition was very unified with respect to vouchers. "All these groups that are in my coalition never worked together. I think it's the only one issue they could all agree on." Another member of the coalition agreed, stating "we basically came to a consensus. We have some very differing views, but when it comes to things like vouchers and that whole debate, historically we put our differences aside until that battle is over." Opposition to vouchers "is one [issue] we agree on whole-heartedly."

Several respondents indicated that charter schools were a political compromise between the pro- and anti-voucher coalitions. A senior staff member for a Hispanic state representative said charter schools were a compromise "because various interest groups could not get vouchers." A member of one of the teacher groups agreed, stating,

> there was some compromising going on because both of those things [vouchers and charters] were going on at the same time. You have to recognize that there is going to be some give-and-take going on and I think that is what happened with charters. There could have been more opposition to charters had not the attention been [diverted to vouchers]. You just have a limited amount of time that you can address each one.

A member of the pro-voucher coalition agreed, stating, "The teachers unions did not expend nearly as much energy in trying to defeat the charter school expansion that we worked on. The debate over vouchers tended to pull away the expenses and energy of people who are opposed to educational reforms like this. There is something to be said about having a voucher debate and a charter debate at the same time. People tend to expend their energy on the things they see as more damaging and hurtful to the public education system."

A member of another anti-voucher group stated in an interview, "It was . . . a compromise between the teacher groups and the pro-voucher groups. We're saying, 'OK, we will leave you alone and we will call off the dogs on vouchers. We will stop all the rhetoric and you go ahead with these charter schools.' We thought, you know, it's only going to be twenty. It's a pilot. You go in as a lobbyist and I hate to say make a deal, that sounds just so . . . but basically that's what you do."

Several respondents suggested that the anti-voucher coalition conceded on charter schools as part of a strategy to "take some of the steam out of the voucher movement."[84] A spokesperson for CPS stated, "we knew that it could be, you know, the people promoting vouchers are saying 'We need more choice' and we thought, well, charter schools could provide a little more choice." A former staff member of the Senate Education Committee observed that "charters were seen as a middle ground between those who favored a pure voucher system where public money would be used as a way by which students could get their tuition paid at a private school or a private parochial school . . . as opposed to those people who absolutely feel that public education had to be controlled by a public entity, i.e., a school board and the state." According to an official with the Texas

Education Agency,

> My impression was that we got a charter bill because the more moderate factions panicked that if they didn't do something, we would end up with a voucher bill. So even though the voucher, the groups that are pushing the vouchers are helping us [charter school advocates], they see it as vouchers are the next logical step after charters. The other side is supporting us because they think that if charters are successful, they'll be no need for vouchers . . . Both sides see us as positive but for different reasons. Choice people, the really heavy choice people say, "Well, we're done with charters, now we can push right on and do vouchers." And the other group is saying, "Man, we have a real strong charter system, we don't need to go into vouchers."

A state senator agreed, stating, "I think there is the possibility that they [anti-voucher groups] thought if we'll let charter schools take hold, the pressure will come off for vouchers and frankly I think they are right . . . There are a lot of people who would say, 'Well, look, students have a choice.'" However, at least one state representative thought the political compromise would backfire and would accelerate movement toward a voucher system. He stated, "By endorsing charters, they [groups opposed to vouchers] are hastening the end of the education system as a monopoly, a government-run monopoly." Accepting charter schools makes it easier to expand into other forms of choice such as vouchers.

A key committee staff member attributed the defeat of vouchers to unified opposition from the traditional, often fractured public education establishment. The very idea of public vouchers for private and parochial schools had a galvanizing effect on the opposition. He noted that the "public school lobby—a lot of those people don't normally agree about anything. I remember telling Senator Bivins during the session, 'Congratulations. You have managed to unite the entire school lobby which no one has ever been able to do.'"

Other participants, however, believe the pro- and anti-voucher coalitions are in a state of flux, with the pro-voucher coalition gaining momentum. A representative for one of the teacher groups stated:

> Most of the African American (legislators) in the House are still with the coalition [CPS] on this issue. There is some grass-roots efforts out there where . . . of course, if you look at inner-city schools and they're predominately African American and Hispanic, so they're feeling the brunt of a lot of this in terms of schools falling apart, year after year they are on the list of low-performing schools. At some point you sort

> of get tired of that so the pendulum is definitely swinging [toward vouchers]. You know. It may be sort of a slow swing but it is definitely swinging.

An official for one of the teacher's groups concurred, noting that legislative voting patterns are slowly changing, with an increasing number of Democrats willing to try pilot voucher plans. She stated, "Republicans tend to favor it more but it's not, the lines have grayed considerably. I've gone to forums and listened to people talk about vouchers and it's gotten to that point where the lines as far as ideology are gray. People say, you know, I mean, 'these schools are falling apart, kids aren't learning. I want out.' The lines are certainly blurred."

One longtime staff member with extensive knowledge of the politics behind the voucher battle commented, "What I think is happening, though, that makes this a little more complex than just a simple partisan analysis and also favors, while we're seeing this shift in momentum toward vouchers, is that you're increasingly seeing more minorities gravitate toward the pro-voucher position. I think the ability to stop the momentum toward vouchers is increasingly eroding politically." A Republican state representative who strongly supports vouchers agreed, noting:

> We are [getting bipartisan support for vouchers], especially with minorities who represent inner city kids or have the worst conditions and with other liberal Democrats. Last time we got the chairman of public ed. [the Public Education Committee], Paul Sadler (D), voted for vouchers last session. I know that's a first in Texas history, that may be a first in U.S. history, to have a Democratically controlled House and have the chairman of the public ed. committee vote for vouchers.

Fissures and Fractures in Voucher Coalitions

Unfortunately for supporters of vouchers in Texas, not long after joining the pro-voucher coalition, Lieutenant Governor Bullock resigned his position from the group, citing the intense partisan politics of the coalition.[85] Mansour, head of Putting Children First, wrote a letter in January 1998 seeking campaign money to unseat Democratic legislators in the upcoming fall elections, including replacing House Speaker Pete Laney, a voucher foe, with a pro-voucher Republican.[86] In the letter, Mansour wrote, "We have everything in place to achieve a victory for school-choice advocates across the nation, and I refuse to allow one or two individuals at the Capitol to block our efforts."[87] In a letter to Mansour, Bullock responded, "I was assured that when I accepted the honorary chairmanship of Putting Children First that it would be a bipartisan effort to help Texas kids and improve education in Texas. Political partisanship has brought virtual gridlock to the deliberations of the United States Congress and, from time to time, threatens to hinder genuine progress in the Texas Legislature." Bullock indicated that he remained supportive of a limited pilot voucher plan.[88]

Another senior staff member commented upon differences within the coalition supporting vouchers and among members of the legislature sympathetic to their cause. She noted that some of the voucher people "are very ideological" and want a wide open voucher plan, but neither Senators Ratliff nor Bivins will support that. She asserted that "we had probably as much trouble with them [the pro-voucher groups] as we had with the anti-voucher people." She noted that Senators Bivins and Ratliff said "we don't know whether vouchers will work or whether they are the answer but we ought at least allow the pilot program for students in poor performing schools to do this." However, "for a lot of the people in the pro-voucher front, that's not near enough." She believes that some of the pro-voucher groups lack political savvy despite being connected to a national political movement. A state representative agreed, stating, "They [the pro-voucher coalition] made some incredibly stupid decisions last cycle of elections. They went into Republican primary contests where you pretty much knew who was going to win and they gave against the winner in big amounts because they thought the other person was more pro-voucher. Which is true but it's not very pragmatic." He observed that they gave $10,000 against Terry Keel and "they gave $10,000 against Dennis Bonnen. Both of those people ended up voting against vouchers last session and specifically did so to send a message to that PAC [political action committee]. Neither one of them are against vouchers. They are both pro-voucher. They were sending a message."

With these tactical errors made by the pro-voucher coalition, and in the presence of committed, vocal opposition by a coalition composed of the state's major teacher organizations, professional administrator associations, the Texas PTA, the Coalition for Public Schools, and the Texas Hispanic Families Coalition,[89] voucher plans introduced in each of the last six legislative sessions failed (1991–2001), despite the support of then-Governor Bush and current Governor Rick Perry. In response to such strident opposition, pro-voucher supporters have had to alter their strategy to craft proposed voucher bills as pilot programs, thereby attempting to maximize their appeal to ethnic minorities among Democratic party legislators.[90] As the 1999 legislative session drew to a close, a solid block of 11 Democrats, aided by a handful of anti-voucher Republicans, effectively blocked S.B. 10, a proposal to create a pilot voucher plan, from coming to the Senate floor for debate.

In response, the lieutenant governor attached riders to two bills, HB 2307 and HB 3765, which were to be scheduled for Senate debate.[91] In April, with fewer than 40 days remaining in the legislative session, Lieutenant Governor Rick Perry made major modifications in S.B. 10 to "make it more palatable," including paring the number of eligible school districts down to two but allowing children from families of all income levels to apply for vouchers.[92] Anti-voucher Republicans were repeatedly called into meetings with the lieutenant governor, and Governor Bush made several rare appearances in an effort to persuade senators to allow a voucher bill to come to the floor for debate and vote.[93] However, despite personal appeals and creative attempts to attach voucher proposals as riders to other legislation, voucher plans have been consistently blocked in the Senate.

Summary

The pro- and anti-voucher coalitions are easily identifiable and well entrenched in state-level political battles over school choice. The differences between the coalitions are clear and differences within each coalition are evident. Both coalitions use a variety of influence and pressure tactics to achieve their political objectives. Reflecting its intense political divisiveness as a radical strategy of educational reform, vouchers draw both supporters and opponents into the front lines of battles over school choice. The issue pushes people to extremes, with little common ground or room for compromise. Even small, pilot voucher plans are opposed with intense fervor by the traditional public education establishment.

It is also clear that key tactical mistakes by the pro-voucher coalition has hurt the prospects of passing voucher legislation in Texas. A senior staff member of one of the education committees suggested that the pro-voucher coalition gave legislators sympathetic to their cause nearly as much trouble as the anti-voucher coalition. The staffer cited the ideological nature of many members of the coalition and their insistence on a wide open voucher bill, which the staffer said would never get through the legislature. The pro-voucher coalition engaged in such damaging efforts as contributing money in Republican party primaries to candidates who were more intensely supportive of vouchers than the incumbents.[94] This led two of the pro-voucher incumbents to vote against voucher bills supported by the coalition during the 1997 legislative session. The coalition also spent a great deal of money trying to defeat House Speaker Laney when he ran for reelection. In Laney's case, he went from being neutral toward vouchers to actively opposing them. Given his position as speaker and ability to influence legislation, the efforts of the pro-voucher coalition were a major tactical blunder.

Charter Schools, Interest Groups, and Advocacy Coalitions

What is most remarkable about the politics surrounding charter schools in Texas and in many other states is how little opposition has arisen to various charter school proposals, in comparison to vouchers. In an address to the Texas Parents and Teachers Association, state Representative Wilhelmina Delco said that Texas was following the national trend by creating programs allowing for greater parental choice.[95] Then-Governor Ann Richards and Lieutenant Governor Bob Bullock, both Democrats, supported charter schools "as a way to give parents choices about their children's education without diverting public dollars to private schools."[96] Sandy Kibby, legislative consultant for the Texas PTA, said that Bullock's proposal for charter schools was "in line with the Texas PTA's position on school choice."[97] Republican state Senator Bill Ratliff, then-chair of the Senate Education Committee, also supported the creation of charter schools.[98] In testimony before the Senate Education Committee, Ed Adams of Texans for Education, a coalition of major Texas employers working to improve Texas public schools through sound public policy, stated that his organization advocates the creation of charter schools as a vehicle for education reform within public education.

Rallying support in the legislature, Republican state Representative Kent Grusendorf asserted charter schools "give more latitude to parents and schools in helping their students achieve. They are positive changes."[99] A senior official at the Texas Education Agency commented that the Republican Party is a big advocate of charter schools in Texas. In an interview, she noted that "the biggest supporter we [advocates of charter schools] have period in the state of Texas, bar none, is Governor George Bush. He's our number one proponent." She said, "the Governor's Business Council is probably the single organization that has had the most influence [advocating for charter schools]."

The charter schools movement was given a significant boost when several of the state's major public school advocacy groups threw their support behind the effort. In Texas, none of the major teacher, administrator, or parent groups took a stance in opposition to charter schools. For example, in a position statement, the TCTA supported a broad range of school choice within the public school system but opposed vouchers, tax credits, and other proposals that would have permitted the use of public funds for private schools.[100] Billy Walker, executive director of the Texas Association of School Boards (TASB), stated that his organization stands behind charter schools "as long as local boards are driving the bus."[101]

However, John O'Sullivan of the Texas Federation of Teachers was cautious about proposals for creating charter schools, arguing that it was unclear exactly how such schools would be constituted and what state rules and regulations would apply to them. This concern about what charter schools would look like and what state regulations they would be exempt from is where opposition to various charter proposals arose. In an interview with an official with the Texas Federation of Teachers, the representative said the TFT was not opposed to the concept of charter schools if they were organized by teachers or in conjunction with teachers, but he was suspicious of efforts not inclusive of teachers. In testimony before the Senate Education Committee, Brad Duggan, executive director of TEPSA, expressed support for charter schools and home-rule charter schools but was opposed to eliminating the 22 : 1 limit on class size. An official at the state education agency noted that many teacher groups "are not [fully] supportive of the charter movement because the charter movement is allowed to hire people who are not certified and they feel like that is inappropriate. And about half the charter teachers are not certified."

In Texas, divisions within the traditional Education Establishment that had been papered over in the debates over vouchers became readily apparent in debates over charter schools. In an interview, a senior official with TEPSA said "some of the teachers groups are adamantly going to oppose the independent charters because there is no requirement that teachers be certified. We're going to take a more compromising position." Testifying on behalf of TASB, Melissa Knippa said her organization supports home-rule charter schools and advocates less state regulation of public schools. She testified, "This home-rule school concept is a good first step in the direction of deregulating our public schools. We believe that our home-rule school districts should be free from everything except statewide accountability system requirements, school finance requirements, federal and constitutional provisions." Virginia Collier, legislative chair of the Texas Association of School Administrators (TASA) and herself a school superintendent, testified in support of home-rule charter schools. She said, "we believe that the home-rule charters can free local districts from state-imposed rules, regulations, and mandates."

However, TSTA President Richard Kouri criticized Bush's plan for home-rule charters, stating "the way I understand home rule, it would eliminate all the mandates out there. Local school districts would be given total discretion about how to handle schools," thus raising the specter that schools could lower passing requirements to benefit certain student groups such as athletes.[102] Kouri stated "what we see in that proposal is doing away with all state requirements and saying that local districts can make all of the decisions regarding their schools" and expressed fears of a return to overcrowded classrooms and overworked teachers.[103] In testimony before the Senate Education Committee, Kouri pointed out that home-rule charter schools do not have to take students with disciplinary or criminal histories. He stated, "if they [charter schools] are going to be public schools, then they need to serve all possible students, not just model students." However, several officials noted that although proposals for charter schools would free up schools from many state laws and regulations, several important laws and regulations would not be waived, including school finance requirements, testing and accountability, bilingual and special education programs, no-pass, no-play, maintenance of student-teacher ratios through the lower grades, as well as federal rules and regulations.[104]

As a compromise measure, several restrictions were placed on charter schools in S.B. 1. A senior staff member of one of the education committees explained:

> The way the law [charter schools] came out, it was real restrictive. There were still certain restrictions that were placed on it by TEA [Texas Education Agency]. In other words, it was still governed under the TEA umbrella. You had to apply to the State Board of Education. As part of the application you had to demonstrate that you were still going to do certain things that tied the charter to the public school system. For example, bilingual, special ed., nondiscrimination, adhering to the state's academic accountability system, allowing teachers to participate in TRS [Teacher's Retirement System]. There were certain protections in there that I think made people feel comfortable that you weren't gonna get a, ah, no Branch Davidian-type person essentially opening up a school saying, "We're ready to provide educational services."

A member of TCTA observed during an interview that "the positions of the individual organizations [on charter schools] are different enough so that it is not possible" to adopt a unified position. In an interview, a member of TEPSA concurred, stating "the associations in Austin form a lot of coalitions. Even within those coalitions, there are times [such as charters] when we have to agree to disagree." These views are consistent with that of a senior staff member of the Senate Education Committee who stated:

> CPS [the Coalition for Public Schools] was less united in their opposition to charters. I think the reason charter schools got through in S.B. 1 was because they were part of a huge package. And there were things in there that the teachers' groups liked—like tying the salary schedule to total state funding, raising it to twenty steps instead of ten, protected their contract rights. And had that been a separate bill, I think it [charters] might have had problems getting through. But by putting everything in there, like Senator Ratliff said "there is something in here for everybody to hate" but it's all one package and we are going to go through with it. Plus, it came down a lot from what was originally proposed. Originally, there were no limits [on the number of charters that could be granted by the State Board of Education. The final bill created a pilot program of only twenty]. So, it was very much a compromise.

Summary

It is clear that charter schools in Texas won legislative approval because (1) charter schools had widespread bipartisan support in the legislature; (2) they were supported by both Democratic and Republican governors; and (3) what little opposition there was to charter schools was not united into a coherent opposition advocacy coalition. Various charter school proposals did not generate the enmity that arose in opposition to vouchers. The evidence suggests that the current status of school choice in Texas (as written in existing legislation) is a product of the competition between advocacy groups.

To date, the anti-voucher groups have been able to prevent passage of vouchers for use in private or parochial schools, while the pro-voucher coalition has been successful at creating a system of open choice within the public school system. These findings support the view that the legislature's adoption of charter schools and rejection of vouchers was the product of competition among advocacy coalitions, as research by Sabatier, Jenkins-Smith, and others would suggest.[105]

The legislative battles over charter schools and vouchers provided an ideal setting from which to view these competing coalitions in action. Charter schools garner bipartisan support because they offer "safe" choice,[106] freedom from some state regulations, and greater accountability[107]—all the while retaining their public character, including provisions of open enrollment, nondiscrimination, and remaining tuition-free.[108]

With respect to charter schools, the lack of unified opposition was a major factor in getting the reform through the legislature. All the members of the pro-voucher coalition supported the creation of charter schools. In addition, nearly all members of the major anti-voucher coalition, CPS, advocated the creation of charter schools; in fact, a few of the members actually assisted in the creation and operation of these schools. Public school advocacy groups supportive of charter schools included the Texas PTA, a bipartisan group of legislators (Speaker Laney and Lieutenant Governor Bullock among them), Governor Richards and Governor Bush, the Governor's Business Council, the Coalition for Quality Education,[109] Texans for Education,[110] the major teacher and administrator associations, and numerous other organizations, although significant differences were evident in the type of charter schools each group supported. In each of the last three legislative sessions since the charter school law was passed, concerted attempts have been made by charter school supporters to remove many of the regulations (particularly accountability regulations) governing charter schools. However, these attempts have been unsuccessful.

SCHOOL CHOICE AND ADVOCACY COALITIONS IN COMPARATIVE PERSPECTIVE

Charter Schools

Joseph Viteritti, professor of public policy and education at New York University, observed that, "Broad political coalitions composed of African-American parents, white liberals, urban Democrats, and business leaders, as well as market-oriented conservatives and Republicans" have joined forces to support various school choice initiatives.[111] Advocacy coalition activism in school choice in Texas appears consistent with research by Paul Bauman, Chris Pipho, and others who found that in many states and communities, charter schools have bipartisan support.[112] President Bush and former President Clinton support charter schools, as do conservatives and some members of the teachers' unions.[113] The National Education Association (NEA) has even launched some charter schools of its own.[114] In several states, including New York and Michigan, Republican governors were instrumental in securing legislative passage of charter school legislation.

In New York, Governor George Pataki effectively held the state legislature hostage, threatening to veto a significant legislator pay raise unless he was presented with a strong charter school bill.[115] Reporters observed that, "In the finest give-and-get tradition of Albany, Mr. Pataki made it clear that he would sign the pay measure [the first raise in a decade] only if the Legislature withdrew its long-standing opposition to charter schools."[116] In several other states, Democratic governors were key players in the adoption of charter school legislation. Colorado, Georgia, New Mexico, and Missouri initiated charter school reforms while having a Democratic governor.[117] Thirteen states that passed charter school legislation had legislatures controlled by the Democratic Party.[118] A bipartisan coalition in Colorado was instrumental in pushing that state's charter school bill through the legislature.[119] Democratic Governor Roy Romer, a staunch supporter of the legislation, provided the strong leadership necessary to foster "the bipartisan coalitions that developed in support of the legislation."[120]

In Ohio, proposed charter school legislation was opposed by several members of the traditional education establishment. The Ohio Education Association[121] opposed the measure, "fearing the community schools would siphon money from an already underfunded school system."[122] The State Board of Education expressed concern over "which public authorities would sponsor and monitor the new schools."[123] However, several organizations came together in Ohio to form an advocacy coalition supporting charter schools. One such organization—We Can—consisted of a racially mixed group of 22 Catholic and Protestant churches working to promote greater parental involvement in education using community schools (as charter schools are known in Ohio) as a vehicle to do so.[124]

However, in many other states, the Catholic Church does not enthusiastically support charter schools, since in many urban areas, parochial schools report losing students to charter schools. In urban areas where the majority of charter schools are concentrated, many parents who are not Catholic choose to send their children to parochial schools because they believe they are smaller, safer, and offer an enriched academic environment—benefits that charter schools also offer. With the expansion of charter schools in urban areas, many parents are selecting the "free alternative" rather than the more expensive parochial school option.

Opposition to charter schools in Ohio was much more substantial than in Texas, in part because Cleveland's pilot voucher plan was created before the Ohio legislature passed charter school legislation. On the surface, with passage of the voucher plan, one would think that expanding into "safer choice"—public school choice—would have been easy. As an editorial writer for the Cleveland *Plain Dealer* noted, "Go figure: Ohio boasts one of the most controversial school choice measures in the country, yet can't bring itself to pass a modest provision that nearly two dozen other states have already adopted."[125] This oddity is explainable, in large measure, precisely because the state legislature had already passed a voucher bill. The threat of vouchers no longer existed—it was reality. Accordingly, the traditional public education establishment in Ohio could focus all their energy, attention, and resources on opposition to charter schools.

Interestingly, however, the issue of mayoral control played a role in diffusing union opposition to charter schools in Ohio. In 1996, House Education Committee Chair Michael Fox, a Republican, made it known that state legislators were ready to turn over control of the Cleveland public schools to Mayor Michael White—a move which the Cleveland Teachers' Union opposed. Fox suggested that if union leaders softened their opposition to school choice, then he "might just have ammunition enough to slow the mayoral bill's momentum."[126] He recommended that groups opposed to mayoral control of schools should form a coalition to help pass a charter school law.[127] Fox stated, "The political imperative of giving it to the mayor is irresistible unless there's an alternative that had big support from stakeholders. They're [teachers' unions] not saying they're for [charter schools], but they're saying maybe. What was significantly different is that in past conversations with those guys, the answer was simply 'no way.' "[128] Richard DeColibus, president of the Cleveland Teachers Union, remarked after several meetings with Fox, "If it comes down to that choice, it might be a matter of figuring out which is the least of two evils. We aren't inherently against charter schools, as long as teachers aren't penalized in the process."[129] Another decisive factor leading to passage of Ohio's charter school legislation in 1997 was the deal struck between the Ohio Education Association and members of the legislature whereby the union agreed to drop its opposition to the reform in exchange for a provision allowing teachers "to keep their contracts in schools that switch to the charter format."[130]

In Pennsylvania, despite wide bipartisan consensus (a charter school bill passed the House by unanimous vote [199-0] on March 13, 1996), Governor Ridge threatened to veto it—on the grounds that it did not give enough flexibility to those who wanted to create their own schools.[131] In effect, the governor believed the bill reflected too much the views of the Education Establishment, particularly the teachers' unions and state school boards' association. Furthermore, some voucher proponents were not enthusiastic supporters of charter schools when the legislature was debating charters. Voucher supporters said "they were fearful that if the Legislature approved a charter schools bill, lawmakers would be less inclined to consider tuition vouchers," which proponents believed to be "a more 'fundamental' and significant public school reform."[132] However, state Education Secretary Eugene Hickok contended that "approval of a charter schools bill would make it easier to win approval of more fundamental reforms, such as tuition vouchers."[133]

One reason for the lack of strident opposition to charter schools in many states is that "school choice has been used in many forms ranging from alternative education programs, magnet programs, inter- and intradistrict transfers, tax credits and vouchers that can be applied toward tuition and other educational services."[134] In Texas, proposals for charter schools, including the granting of home-rule charters, are similar to the home-rule charters granted to Texas cities with populations of 5,000 or more, which allow home-rule cities to adopt any form of governance they choose.[135] This is consistent with former state Education Commissioner Lionel "Skip" Meno's observation that some districts already allow intra-district choice options.[136] An analysis conducted in 1994 by the House Research Organization found little in Texas state law that prevents districts from creating open enrollment charter schools or privatizing various school services and programs.[137] Accordingly, the legislature's formal incorporation of charter schools into S.B. 1 was not viewed by many in the traditional education establishment as a radical or threatening type of reform.

This finding is consistent with research that concludes that the major political advantage of charter schools is that they occupy something of a middle ground between the public education system as it is currently structured on the one hand and a voucher system on the other.[138] Some researchers describe charter schools as "schools that are privately run but publicly regulated and financed, a system of governance intended to promote both autonomy and accountability."[139] Since they do not go beyond the rhetoric of the public sphere, the scope of the conflict over charter schools is accordingly limited.[140] As such, charter schools are more politically palatable to state legislatures committed to educational reform because charter schools do not represent a significant departure from current practice (after all, at their core, charter schools reflect the central tenets of public school decentralization).

A growing body of research suggests that charter schools were a political compromise, "an attractive alternative to doing nothing to improve the public schools or abandoning them altogether."[141] In Minnesota in 1985, Republican Governor Rudy Perpich encouraged the adoption of a statewide interdistrict open-enrollment choice plan allowing students to cross "district lines to attend any school where space is available" after a voucher plan failed to get through the state legislature.[142] Although the initial choice legislation was narrowly defeated in 1985, it passed two years later, largely due to persistent pressure from Governor Perpich.[143] In North Carolina, charter schools were accepted as a less-threatening alternative to tuition tax credits.[144] In Arizona, vouchers were used "as a red herring" to "distract unions from focusing on their opposition to charter schools."[145] The strategy was to threaten the teachers' unions with vouchers, thereby making "them willing to contemplate compromise [charter school] legislation."[146]

In California, Democrats gave their support to charter schools in the hope that "this variant of choice would diminish support for Proposition 174, the voucher measure that was about to be put to the voters."[147] The voucher initiative, led by the chair of the California Business Roundtable and generally supported by the business community, would have allocated about $2,600 per child for use at a public or private school.[148] The California Education Association spent more than $12 million to defeat Proposition 174, with 70 percent of voters ultimately against the measure.[149] In an effort to take the steam out of the growing school choice movement in California, the state legislature passed the Charter Schools Act of 1992, sponsored by Democratic state Senator Gary Hart.[150] The repercussions of the voucher battle, coupled with opponents' fears that the measure would be resurrected, were so real that, "The California legislature also passed a pair of public school choice bills in 1994 patterned after the Minnesota open-enrollment legislation. One bill authorized school districts to allow interdistrict transfers, provided that classroom space is available and the transfer does not disturb the racial balance of the schools. The other bill requires school districts to allow intradistrict school transfers, provided that there is space in the desired school."[151]

The timing of the legislative proposals—with charter schools and vouchers being considered during the same period—facilitated passage of the less controversial legislation.

In Texas, the adoption of charter schools over vouchers was a compromise between competing advocacy coalitions. Comments by policymakers such as "give and take," "make a deal," "take some of the steam out," and "take the pressure off" were common. This finding is consistent with research conducted in other states and at the national level. Bruce Cooper and Vance Randall found that to take the steam out of the voucher threat, opponents offered a host of "choice" alternatives including magnet schools, charter schools, school-based decisionmaking, and "out-sourcing" or contracting schools out to private, for-profit firms such as the Edison Project.[152] According to the researchers, "These mild forms of choice were release valves for pressures to seek radical privatization—making changes at the margins but keeping the public school system firmly in control and fundamentally unchanged."[153] In many states, charter schools receive bipartisan support because, "Opponents of vouchers pin their hopes on charter schools as the public alternative to private school aid, even as voucher advocates support charters as a foot in the door of the public school monopoly."[154]

In many states, opposition to charter schools arose from local school boards, teachers' unions, and school district administrators (especially superintendents) who feared losing power, students, and money to new charter schools. The National School Boards Association (NSBA) lobbies to give only local school boards the power to grant and renew charters—a position vehemently opposed by charter school advocates such as the Center for Education Reform, which argues (correctly, as it has turned out) that concentrating charter-granting authority in local boards will severely limit the number of charters granted in a state. In New York, charter schools passed despite strident opposition from local school boards and the state's two powerful teachers' unions. In New York City, then-Chancellor Rudy Crew publicly attacked New York's charter school law, arguing that it would drain resources from the rest of the public school system. Crew later attempted to control the charter school movement in New York City by announcing the creation of four new charter schools, the conversion of six existing schools to charter schools, and permitting two community school districts in Brooklyn to experiment with local control.[155] The Chancellor candidly stated that he "wanted to prevent teachers and students from leaving the city school system and enrolling in charter schools that he did not control."[156]

Despite the enthusiastic support of Al Shanker, late President of the American Federation of Teachers, teachers' unions throughout the country have expressed concern over (1) teacher certification requirements, (2) waiving of collective bargaining agreements, and (3) employment of non-union teachers. As Joseph Viteritti observed, "Many of the state and local regulations from which charter advocates sought relief had originally been drafted at the behest of school boards and teachers unions."[157] Several researchers note that "the lion's share of their [the NEA and AFT] interaction with charters remains confrontational at the state and local levels."[158] Even as the teachers' unions "launch their own charter pilot projects and laud these schools as laboratories of innovation, they press for the rejection or weakening of charter bills in one state after another."[159] The Center for Education Reform refers to this as the "death of a thousand cuts" political strategy.[160]

In Texas, the intense pressure brought to bear by voucher proponents—the very real threat that a voucher bill would pass the legislature—muted any real opposition to charter schools and significantly enhanced their attractiveness and chances for legislative passage. This finding is consistent with how the legislative dynamics played out in other states such as California and Arizona, where charter legislation would probably not have passed (or, if it did, would not be nearly as strong) "without the threat of vouchers to soften the opposition."[161] In both states, "charter schools were initially regarded as a way to forestall vouchers."[162]

However, not all proponents of school choice endorse the charter school movement with equal enthusiasm. As Michael Mintrom and David Plank point out, charter schools not only attract students from neighboring public schools, but from private and parochial schools, and home schooling, as well.[163] In several cities across the country, "charter schools are posing new competition for Catholic schools by expanding the educational options for urban families."[164] Although data on student transfers from private and parochial schools to charter schools are not collected at the national level, some dioceses such as Newark and Detroit have begun to track it in their respective regions—which suggests that losses are large enough for Catholic leaders to be concerned and take notice.[165]

In several states, significant political battles are being waged either to expand charter schools (for example, by lifting the cap) or to restrict the scope of charter school legislation (for example, by imposing greater regulation and state oversight). In states that place a cap on the number of charter schools, each legislative session inaugurates another round in the cap wars. In 1999, Governor John Engler's proposal to raise the cap in Michigan was blocked by a solid wall of bipartisan opposition—including many moderate Republicans representing "suburban school districts with excellent public schools, whose citizens have little to gain from expanded school choice."[166] A similar measure failed in North Carolina two years later.[167] In New York, where the governor literally strong-armed the legislature into passing a charter school law, no fewer than 17 bills were pending in the legislature that would curb or restrict charter schools, including proposals to subject charter schools to school board review, requiring teacher collective bargaining, reducing charter school funding, and prohibiting for-profit charter schools.[168]

The Illinois legislature recently considered increasing the limit on charter schools in Chicago, while at the same time imposing greater limits on the number of campuses on which charter schools can operate, excluding for-profit companies, and increasing the percentage of certified teachers in charter schools to at least 75 percent.[169] On the other hand, the California legislature "significantly liberalized its charter school law" in response to the threat by a group of Silicon Valley businessmen who "gathered enough signatures to demonstrate that it could place an even more open-ended charter school initiative on the November 1998 ballot."[170]

Efforts are underway in several states to impose greater state oversight on charter schools (both in the process of granting charters and in the operation and accountability of the schools themselves). In Hawaii, an independent review panel was created to oversee charter formation.[171] In some respects, this legislative activity is not surprising, as states make refinements to existing legislation. Considerable variation exists in charter school legislation from state to state. For example, Arizona, which has a wide open charter school law, does not require background checks and fingerprinting of teachers in charter schools—although teachers in regular public schools must undergo this process. In 1999, the California legislature, under pressure from the teachers' union, considered mandating collective bargaining for charter schools. In Ohio, the legislature imposed a limit on the number of charter schools operating in the state.[172] In New Hampshire, Governor Jeanne Shaheen vetoed a bill that would have substantially strengthened the state's charter school law and allow multiple authorizers, rather than first having to go through the local school district.[173]

Vouchers

Unlike charter schools, which are far more likely to engender wide bipartisan support, and, therefore, create less readily identifiable advocacy coalitions, the politically charged nature of school vouchers tends to create powerful and persistent coalitions (both for and against). However, these coalitions tend to differ somewhat in composition in each state. The Black Alliance for Educational Options (BAEO) consists of a coalition of more than 700 pro-choice activists in 35 states. A poll by the Wisconsin Policy Research Institute indicates that a significant majority of African American respondents in Milwaukee support vouchers.[174] In Denver, African American and Hispanic parents filed a class-action lawsuit in an attempt to get the legislature to create a voucher plan for low-income, predominately minority parents.[175] In Michigan, key voucher supporters include the Catholic Church, the Chamber of Commerce, and some leaders of the African American community in Detroit.[176] Looking at voucher coalitions across the 50 states, support for vouchers is clearly no longer a Republican issue, if it ever was.[177] Supporters of vouchers now include such liberal groups as the Annie C. Casey Foundation and The Urban Institute.[178]

A significant difference in advocacy coalition activism in Texas compared to other states is the presence of organized opposition to vouchers by a broad-based coalition of public school advocacy groups, institutionalized through the creation and operation of CPS, whose sole mission is to defeat voucher legislation. This is radically different from the situation in Milwaukee, where, according to Jim Carl, "public school advocacy organizations, composed of middle-class reformers and researchers, did not move to oppose the Milwaukee proposals."[179] It was in this vacuum that choice plans came to fruition. In New York, a state with incredibly strong teachers' unions vehemently opposed to vouchers, "nearly every member of the political establishment...opposes private school vouchers for poor children while refusing to send their own children to public schools."[180] In effect, "the political and economic elite of New York views public schools as places for other people's children."[181] In a strongly worded, 10-page pastoral letter, Cardinal Edward Egan and other New York bishops criticized the state's teachers' unions for their support of the status quo and opposition to vouchers. Cardinal Egan stated, "They [teachers' unions] influence legislators not only through millions of dollars of direct contributions to political campaigns each year, but also through their own political activities designed to elect legislators who will support them and defeat those who will not."[182]

Apart from differences in the institutional context, the voucher movement in Texas is significantly different from efforts in other states. Unlike the coalition of neoliberal and neoconservative reformers who pushed through parental choice in Milwaukee, no broad-based political coalition pushed vouchers in Texas. According to Carl, "In the state legislature, black Democratic and white Republican interests dovetailed with parental choice legislation. Bipartisan support for vouchers helped create a rift among the legislature's white Democrats, with conservatives supporting them and liberals who led the Democratic party in opposition."[183] As a result, "the liberal–labor coalition that had traditionally opposed school vouchers was fatally split," a scenario similar to that which occurred in Cleveland.[184]

However, in Texas, with the notable exceptions of Representatives Wilson and Cuellar, the voucher movement ran into a solid wall of opposition among minority legislators in the state legislature. In Texas, as in California, Oregon, and Colorado, school choice has a relatively fragile, narrow coalition made up of conservative ideologues, the urban poor, and some parents with children already enrolled in private schools.[185] Few state Hispanic and African American legislators support voucher proposals and their support is essential for passing the legislation.[186] Combined, Hispanic and African American legislators constitute nearly one-third of the members of the Texas House of Representatives and nearly one-quarter of the Senate, with the number of Hispanic legislators double that of African Americans in both chambers. However, Hispanics in Texas are not pushing for vouchers. In fact, many Hispanic organizations actively oppose them. Vouchers plans will not be implemented in Texas without support from (or at least without the active opposition of) this large voting bloc.

The voucher issue has exposed significant fissures within the Democratic Party, since most Catholics (the majority of whom are Democrats) support school vouchers. In Milwaukee, some of the most vocal voucher supporters, such as Polly Williams, are people of color. In 1995, as debate raged in the Ohio legislature, Cleveland Councilwoman Fannie Lewis, an African American Democrat, took nearly 300 supporters in six buses to the state capital, urging legislators to support vouchers.[187] Speaking to an eclectic coalition of parents, children, grandparents, clergy, and community activists, Lewis said, "You didn't come down here to beg, you came down to tell people what you want... This is serious business. We ain't playing."[188] However, key Democratic leaders, such as Ohio Senate Minority Leader Ben Espy, who led the fight against school choice, asserted that charter schools and vouchers would "lead to the creation of two separate and unequal school systems, to the detriment of our state's children and its future."[189] He continued, "Vouchers and charter schools will segregate students according to academic ability, behavior, parental involvement and a host of other factors, with public schools continually coming up short."

In Ohio, the initial impetus for vouchers came from a state education committee appointed by Governor George Voinovich.[190] The committee was chaired by David Brennan, a wealthy industrialist who was a major contributor to Governor Voinovich's campaign and whose wife runs a private school.[191] In Ohio politics, business elites play a crucial role in the policymaking process and exert significant pressure on education reform.[192] Several organizations formed advocacy coalitions in opposition to vouchers, including the Ohio Education Association, the Ohio Federation of Teachers, the Ohio PTA, and the ACLU of Ohio. A major advocacy coalition opposing vouchers was Citizens Against Vouchers, a coalition consisting of over 20 education, labor, and community groups.[193] All major education groups in Ohio opposed the reform. The Buckeye School Administrators, a statewide organization of superintendents, voted 9-1 to oppose vouchers.[194]

The legislative battle over vouchers in Pennsylvania has produced some unlikely alliances, including a loose coalition of 41 groups that came together to oppose vouchers.[195] For example, hard-core conservatives who feared government regulation of private and parochial schools joined forces with traditional Democrats to oppose vouchers. On the other hand, some liberal urban Democrats, such as state Representative Dwight Evans and Senator Anthony Williams, favored school vouchers.[196] In 1998, for the first time in its history, the Archdiocese of Philadelphia organized a voter registration drive, ostensibly to elect more school choice advocates to the legislature.[197] Several conservative Republican legislators representing suburban constituents were under heavy pressure to oppose vouchers amidst concern over minority students enrolling in predominately white suburban schools.[198] Opposition also arose from the state's powerful teacher's union. Pennsylvania Governor Tom Ridge said of the teacher's union, "There frankly, probably is no more powerful adversary. They're tough."[199]

Despite this opposition, a voucher proposal nearly passed the Pennsylvania House of Representatives during the 1995 legislative session, falling a mere seven votes short of passage. Partly, the defeat was the result of unwillingness on the governor's part to made deals—trading bridges and highways for votes.[200] Vince Fumo, a powerful Philadelphia Democrat who is pro-voucher but voted against the Ridge plan, was accused by one Republican staff member as "trying to rob the train" in an effort to obtain additional funding for Philadelphia schools and sundry other political perks.[201] Democrats blamed the Ridge administration—claiming that administration officials mistakenly believed they could get the voucher plan passed without any help or without having "to pay out anything."[202]

Adding to this complexity is the fact that in some states, rivalries exist within school choice coalitions.[203] Disagreements and differences among school choice coalitions are common throughout the 50 states, depending on their constituencies. In Michigan, for example, religiously inspired groups such as the TEACH Michigan Education Fund compete with free-market motivated coalitions led by the Mackinac Center for Public Policy over the preferred type of school choice.[204] According to Morken and Formicola, "Neither group sees eye-to-eye on most matters, when everything about them, including their ideology, history, operating styles, and choice programs are so different."[205] Voucher coalitions are diverse and often fragile. Robert Bulman and David Kirp documented the deep divisions in the pro-voucher coalition in Milwaukee between market-oriented conservatives and equity-oriented minorities as the battle for vouchers raged throughout the 1990s.[206]

The sometimes fragile coalitions that form around school choice are very difficult to maintain, posing a particular problem for voucher proponents. In Pennsylvania, for example, Jeffrey Trimbath, executive director of the REACH Alliance[207] (The Road to Educational Achievement through Choice), said "pro-voucher forces have always had to contend with folks on the right with deep suspicions of government intervention as well as traditional foes of school choice, such as teachers' unions and school boards."[208] Sean Duffy, president of the Commonwealth Foundation, a conservative think-tank based in Harrisburg, said voucher supporters have never been able to count on support from the legislature's most conservative lawmakers.[209] To overcome opposition from conservative Republicans, voucher advocates had to court the support of big-city Democrats—some of whom remained opposed to vouchers.[210]

In California, Proposition 38, a proposal to give parents who already send their children to private schools a $4,000 voucher per child, split the private school community—the Association of Christian Schools International (ACSI) supported the measure, but Catholic bishops adopted a neutral position, noting that the proposal lacked special preferences for poor families or children trapped in low-performing public schools.[211] In Michigan, voters rejected a voucher initiative, Proposal 1, that would have provided private school tuition vouchers for children attending low-performing public school districts, despite the support of ACSI, the Michigan Association of Nonpublic Schools, a state Council for American Private Education (CAPE) affiliate, and the state's Catholic bishops.[212] The campaigns surrounding Proposition 38 and Proposal 1 were the most expensive voucher battles to date, with both sides in both states spending collectively nearly $75 million.[213]

Vouchers were not an important policy issue for legislators representing rural areas of Texas, Pennsylvania, and Ohio—an experience common in other, more rural states as well. In fact, some of the most strident opposition to vouchers has come from an unusual coalition of legislators representing rural and suburban areas. A local editorial from Pittsburgh summed up the feelings of many rural residents. Carolyn Hess asked, "How fair is the voucher system for a child in Gold, PA (population, about 200), who has no choice to make in public or private schools? There are no private schools within miles. We always think in terms of metropolitan America."[214] Rural opposition to school choice extends to charter schools as well, since most rural areas have limited public choice options and are more directly threatened with loss of students and funding to charter schools. This explains, in part, why the 11 states that have not passed charter school legislation are predominately rural—Maryland being the lone exception.

In several states, vouchers expose significant rifts within the Republican Party. In Michigan, Republicans have been "profoundly divided over the question of public vouchers, with Governor John Engler opposing a ballot initiative sponsored by the Chamber of Commerce and other party stalwarts."[215] Governor Engler declared that the voucher initiative was "bad public policy."[216] In Pennsylvania, some of the most strident opposition to vouchers comes from rural and suburban Republican legislators, whose constituents stand to gain little from vouchers.[217] A survey conducted by the Commonwealth Foundation found that while 71 percent of respondents in Philadelphia supported vouchers, only 51 percent of respondents in the Philadelphia suburbs supported the reform.[218] California's Proposition 38, organized and supported by a Silicon Valley venture capitalist, was so broad that the measure fractured the coalition supporting vouchers—with several leading Republicans criticizing the measure as both too expansive and too expensive, leading to a splintering of the school choice movement in California. The ballot measure was resoundingly defeated.

As evidence of the uncompromising position of some voucher proponents, some conservative lawmakers said they would not support Pennsylvania Governor Ridge's voucher plan because of the governor's efforts to impose greater state oversight on private and religious-based day-care centers.[219] "Suspicious of government, some of the Legislature's most conservative members long have been skeptical" of Governor Ridge's voucher plan, arguing that it would lead "to greater government regulation of nonpublic schools."[220] Republican state Representative Thomas Armstrong stated, "They insist on trying to impose these regulations on the day-care centers and yet they say they won't impose regulations on the private schools if they get vouchers. I would still be a vote for vouchers if not for this problem."[221] Republican state Representative Sam Rohrer agreed, stating that the fight over day-care regulations is "a perfectly accurate parallel to make. The issue is not whether or not this governor or administration wants to use vouchers to invade private schools. This governor will not be here forever. If you look at what . . . the [Welfare Department] is doing, it's clear that you cannot guarantee that the government won't try to come in and regulate."[222] Governor Ridge himself acknowledged this concern when he indicated his willingness to require assessment tests for students receiving vouchers, in exchange for support from state Democrats.[223] The governor noted that conservative lawmakers might worry that requiring state testing of students receiving vouchers would represent the "camel's nose under the tent." [224]

School Choice and Anti-Catholicism

Sometimes, controversies over school choice, particularly vouchers, get enmeshed in other hot-button controversies—such as the sexual abuse scandals in the Catholic Church. For example, in Pennsylvania, critics of Governor Tom Ridge's voucher proposal cited their concern that public funding sent to parochial schools would be diverted toward "settling court cases involving allegations of pedophilia."[225] Criticizing the lack of fiscal accountability in Governor Ridge's voucher plan, Democratic state Representative Joe Preston stated, "Unfortunately, people in certain religions have been hit hard by an awful lot of lawsuits. I don't want to see our money to be able to go for those different lawsuits for certain people who do not act appropriately."[226] Preston's remarks were cited by some observers as evidence of the "one of the least-often acknowledged factors inflaming opposition to school choice: Anti-Catholicism."[227]

Although the role anti-Catholic sentiment plays in opposition to vouchers is a matter of dispute (Pennsylvania is one of the few states where the specter has been raised), clearly it is an issue in Pennsylvania politics, where anti-Catholic prejudice has a long history, dating from the founding of the Know Nothing Party in New York in 1849. During the 1850s, the Know Nothings were popular throughout the mid-Atlantic states, reflecting "social, religious, and political anxieties caused by rapid social and economic change."[228] Similar to anti-voucher rhetoric that private and parochial schools will undermine the common good and our democratic institutions, supporters of the Know Nothing Party "imagined they were protecting vital American institutions from conspiracies that sought to undermine republican government."[229] Members "regarded the Catholic Church and its hierarchical organization as antithetical to democracy."[230]

Anti-Catholicism remains strong in several areas of the United States, particularly in regions where Catholics constitute a sizable minority of the population.[231] Historian Arthur Schlesinger, Sr., called it "the deepest bias in the history of the American people."[232] Recognized as "America's most persistent prejudice but also its most accepted,"[233] anti-Catholicism has become insidious "precisely because it is not acknowledged, not recognized, not explicitly and self-consciously rejected."[234] Reflecting this anti-Catholic bias in Pennsylvania politics, a 1998 article in the *Pittsburgh Tribune-Review*, one of the state's leading newspapers, decried the fact that 90 percent of all new immigrants to the U.S. were Catholic.[235]

In fact, "professional educators, academics, and the elite of the educational enterprise" have bitterly opposed both freedom of choice to attend parochial schools as well as any expansion of choice into the non-public sphere.[236] Many state constitutions contain "Blaine amendments" that "lay down more prohibitive criteria for separation [of church and state] than those found in the First Amendment" of the U.S. Constiution.[237] Although it is difficult to accurately gauge the extent of anti-Catholic sentiment and the role it plays in opposition to vouchers, the fact that the vast majority of voucher recipients would utilize vouchers in parochial schools (as is the case in Cleveland) suggests that such sentiment cannot be discounted.[238] Some scholars assert that if the primary beneficiaries of vouchers were not Catholic schools, we would "already have a much broader freedom of choice" in American education than currently exists.[239]

Controversies Sidetracking School Choice

To a surprising degree, controversies unrelated to school choice dramatically affect the chances of passing choice legislation. Contributing to the defeat of vouchers in Pennsylvania was a revolt by conservative Republicans who objected to Governor Ridge's stance on family planning and abortion. In June 1995, Governor Ridge eliminated a provision from the state budget that prohibited state money for family planning from going toward groups that advocate or counsel abortions. Thomas Gentzel, spokesperson for the Pennsylvania School Boards Association, said "one reason they are only close [to mustering enough votes for passage] is because of issues totally unrelated to education, such as legislators talking about family planning."[240]

School choice politics in Pennsylvania has also been affected by the legislature's previous experience with state funding for new sports stadiums in Pittsburgh and Philadelphia. In November 1998, the state senate "approved a bill authorizing millions of dollars for new sports stadiums in Pittsburgh and Philadelphia. But the House failed to consider the measure before adjourning for the year. In February, both chambers eventually passed a stadium funding bill. But during the interim, some senators were the subject of withering criticism while House members were immune from the brunt of such criticism. It's not an experience Senate leaders want to put their members through, especially over the long summer legislative recess."[241] As a result of this experience, Senate leaders, in what one observer called "a game of legislative chicken," wanted the House to take up the voucher bill first.[242] Although Senate passage was likely, and would have given the bill momentum, Senate leaders "did not want to expose senators to the political risk of voting for vouchers, unless they were certain House leaders had enough votes to pass the bill."[243] Ultimately, the House did not approve the measure, and the senators were spared from replaying the stadium controversy.

A campaign finance scandal affected the fortunes of charter school legislation in Arizona. In 1994, the same year the state's charter school law was passed, the Arizona Education Association, the state's largest teacher union, "was politically crippled . . . when it was forced to settle with the state attorney general regarding charges that the union had violated campaign finance laws."[244] In Michigan, charter school legislation passed after Governor John Engler "would not accept a school finance reform package without accompanying legislation permitting charter schools."[245]

In Massachusetts, school choice in the form of charter school proposals took place amidst a host of other educational issues including school finance reform (increased funding for poor districts), the standards movement, school governance changes, teacher career ladders, and the proposed elimination of tenure for teachers.[246]

Republican Governor William Weld made it clear to the Democratic-controlled legislature that acceptance of these broad reforms must accompany any increase in school funding. The state's teachers' unions, the Massachusetts Teachers' Association (MTA)[247] and the smaller Massachusetts Federation of Teachers (MFT),[248] opposed several components of the school reform package, including charter schools; the MTA was particularly outspoken in its opposition to the reform.[249]

However, the inclusion of charter schools as part of a broader school reform package facilitated passage of one of the nation's strongest charter school laws in Massachusetts, despite solid Democratic majorities in both chambers of the legislature. Since the teachers' unions, as well as several other groups within the Education Establishment in Massachusetts, were directly affected by the reform package, none of the groups could "afford to devote substantial attention to blocking the charter legislation."[250] In addition, "the 'carrot' of finance reform—and the likelihood that finance reform would not happen without substantial quality reform—made these groups willing to consider compromises they would never have entertained under normal circumstances."[251]

Finally, relationships among key interest groups change over time. For example, during the 1970s, when teacher strikes and collective bargaining were contentious issues, teachers' unions, school board associations, and administrator groups (the traditional Education Establishment) often disagreed vehemently on key education policies. However, key issues during the 1980s and 1990s have brought these groups into a more collaborative, coalitional mode of operation "epitomized by the formation of a broad new coalition to protect and improve the state education funding levels."[252] Education coalitions became common in state politics in the 1980s and 1990s. Coalitions formed around issues of school funding, teacher dismissal policies and certification requirements, and increased graduation requirements.[253] A Pennsylvania lobbyist noted that cooperation between interest groups "ebbs and flows. Historically, we can agree to certain general principles, but when push comes to shove in the final couple of days [before matters come to a vote in the legislature] each organization has its own priorities and works on them."[254] One Pennsylvania lobbyist trenchantly observed, "The best lobbying is coalition lobbying where you agree to form alliances on issues that you can agree on and then you agree to disagree."[255]

CONCLUSION

In this chapter, we examined the principal tenets of neopluralist advocacy coalition theories of policy change and their application to the school choice movement. While the coalitions supporting and opposing various forms of school choice are varied both in composition and in their degree of cohesion, choice is such a polarizing issue in the electorate that diverse groups come together, forming coalitions and often working together for the first time. Working through the institutional structure, coalitions engage in battle on a variety of fronts, buttressed by forces external to the policy subsystem. The unity and strength of these advocacy coalitions play a crucial role in determining the outcome of choice initiatives.

CHAPTER 5

ORGANIZATIONAL LEARNING DYNAMICS: THE UTILITY OF EXPERIENCE?

Perhaps the slow evolution toward greater school choice reflects an understanding among state policymakers that choice works—that it is an improvement over current practice. A number of scholars in the fields of policy and organizational studies believe that policy change and reform are the products of organizational or institutional learning. Organizational learning may be defined as "a process whereby an organization is able to correct its course of action on the basis of information about consequences of its own previous actions and decisions or on the basis of information about the consequences of the actions and decisions of others."[1] Organizational learning reflects a deliberate attempt to adjust policy in the light of past experience and policy-relevant knowledge. One of the earliest works in the field posits that organizational learning:

> occurs when individuals, acting from their images and maps, detect a match or mismatch of outcome to expectation which confirms or disconfirms organizational theory-in-use (defined as the theory of action constructed from observation of actual behavior). In the case of disconfirmation, individuals move from error detection to error correction. Error correction takes the form of inquiry. The learning agents must discover the sources of error—that is, they must attribute error to strategies and assumptions in existing theory-in-use. They must invent new strategies, based on new assumptions, in order to correct error. They must produce new strategies. And they must evaluate and generalize the results of that new action.[2]

Frans Leeuw and Richard Sonnichsen conceptualize organizational learning more simply as "a process in which an organization continually attempts to become competent in taking action, while at the same time reflecting on the action it takes to learn from its present and past efforts."[3] This definition of organizational learning suggests that an organization is an entity with a form of consciousness such that it can both take action and reflect on its actions.

Aside from the definitional ambiguity inherent in the term "organization," there exists the question of "how can we speak of an organization's 'goals' and 'actions' when, paradoxically, only individuals ever decide, act, or function as organizational members?"[4] Mariann Jelinek asserts that organizations have a life of their own—a synergy beyond the sum of their parts and that the actions of successful organizations must be attributable to more than mere serendipity. According to Jelinek, organizational learning is demonstrable "in the sequential application of generalized insights or approaches, which separates it from mere adaptation or routine iteration of what was fortuitously successful in the past."[5] To determine whether organizational learning has occurred, the following questions should be asked: "Did individuals detect an outcome which matched or mismatched the expectations derived from their images and maps of organizational theory-in-use? Did they carry out an inquiry, which yielded discoveries, inventions, and evaluations pertaining to organizational strategies and assumptions? Did members subsequently act from these images and maps so as to carry out new organizational practices?"[6]

The assertion that organizational learning can occur as a result of actions taken by the organization itself or by other organizations implies that organizations have the capacity to evaluate effectively the actions of others. Harald Baldersheim and Per Stava argue that "an organization may be able to learn both from its own experiences and from those of others. It does not have to make all the mistakes itself in order to learn."[7] However, it does not follow that organizational learning will always lead to improvements in policy and practice, since organizations may draw the wrong lessons from the experiences of others. Peter Hall notes that organizational learning "does not necessarily mean that policy becomes better or more efficient as a result of learning. Just as a child can learn bad habits, governments, too, may learn the 'wrong' lessons from a given experience."[8] Unfortunately, if organizational learning leads to the promulgation of less effective public policies, it begs the question of the utility of the concept itself, weakening its applicability as a theoretical construct. In this chapter, I seek to explore the extent of organizational learning within state legislatures—intensely political, public organizations—to determine whether research utilization and organizational learning affect legislative decisionmaking, thus shaping school choice and policy change.

ORGANIZATIONAL LEARNING AND POLITICAL OR POLICY LEARNING

Before exploring the extent of research utilization and organizational learning in school choice, we must first differentiate organizational learning from political or policy learning because the concepts are vastly different, with significant implications for the policy process. In fact, the terms are frequently confused and used interchangeably in the literature, even though they are quite different. As conceptualized in the literature, organizational learning is best viewed as "academic learning"—as what policies work best. However, political learning occurs when legislators, lobbyists, interest groups, and advocacy coalitions learn how best to package and sell reforms—how best to play the political game and achieve the minimum winning coalition necessary to pass legislation. In their book, *Policy Change and Learning*, Paul Sabatier and Hank Jenkins-Smith refer to this process as policy learning, whereby coalitions learn through experience how best to achieve their political objectives. While such learning obviously occurs and is an important part of the political process, it is significantly different from the research on organizational learning. Accordingly, political or policy learning is not included in the discussion of organizational learning contained in this chapter.

POLICY IDEAS, RESEARCH UTILIZATION, AND ORGANIZATIONAL LEARNING

Several researchers assert that models of policy change that focus on institutions, political systems, interest groups, or culture are inadequate because they ignore the role of ideas in promoting and shaping policy change.[9] Hanne Mawhinney asserts that "ideologies, dominant ideas, and policy-specific ideas are powerful influences on policymaking."[10] The propagation of ideas via research is important because it questions existing perspectives, challenges current ideas, provides "alternative cognitive maps," and helps policymakers develop alternative constructions of reality.[11] Over time, "these alternative images of reality can yield new ways of addressing policy problems and new programs and procedures for coping with needs."[12] Through organizational learning processes, policymakers may use research to discover what policies and programs work best and why.

The key analytical question is how these ideas are used to effect policy change.[13] The problem is that it has "proved very difficult to uncover many instances where social research has had a clear and direct effect on policy."[14] Daniel Callahan and Bruce Jennings found that "occasionally the findings of social scientific studies are explicitly drawn upon by policymakers in the formation, implementation, or evaluation of particular policies. More often, the categories and theoretical models of social science provide a general background orientation within which policymakers conceptualize problems and frame policy options."[15] In fact, there is general agreement that the "direct influence of policy research upon decision-making is rare."[16] In most cases, the influence of social science research on policy is "indirect—one small piece in a larger mosaic of politics, bargaining, and compromise."[17] Carol Weiss agrees, noting "even as research on a subject accumulates and as its methodological and conceptual quality improves, it may provide no firmer basis for choice."[18] The results of research "are often so equivocal and knowledge so imperfect that the policymaker... is forced to choose between equally credible but contrasting views and proposals."[19]

This leads to two common conceptualizations of the role of ideas in shaping policy change. First, policy information is "used in an advocacy fashion, i.e., to buttress one's position or to attack an opponent's."[20] In such cases, research "becomes ammunition for the side that finds its conclusions most congenial and supportive. Partisans brandish the evidence in an attempt to neutralize opponents, convince waverers, and bolster supporters."[21] Research supplies policy actors with knowledge to be used for influence and persuasion.[22] This view holds that policy research serves primarily as policy advocacy. Policy analysis is reduced to policy argument.[23]

The second conceptualization of the role of ideas in shaping policy change is through "a process of 'enlightenment' whereby the findings accumulated over time gradually alter decision-makers' perceptions of the seriousness of the problems, the relative importance of different causes, and/or the effects of major policy programs."[24] In her extensive research on the role of research in policymaking, Weiss found that policy actors claim to be influenced by social science research but, "when pressed to give examples, often cite broad generalizations... or social science concepts... Not single findings, one by one, but *ideas* from social science research appear to affect the development of the policy agenda. They [policymakers] draw on it to understand current conditions, the options available for coping with problems, and the limits of the attainable."[25] The use of research to enlighten policymakers suggests that "over time the cumulative effects of scientific findings have an impact on our understanding of interventions in society and thereby affect the kinds of programs introduced to deal with social problems."[26] Research also plays a role in influencing policymakers' perceptions of social problems, particularly how research is used to frame social problems and policy solutions. It is through such processes that ideas shape policy change.

Organizational Learning, Rationality, and Politics

A major criticism of the organizational learning approach for analyzing policy change is similar to criticisms of other rational choice approaches to education reform: that policy solutions in the private sector are not directly transferable or applicable to the public sector. Most of the research on organizational learning is located within the private sector and is based on a commonly accepted principle that private organizations such as businesses are capable of organizational learning. As Jelinek observed in her doctoral research at Harvard, the long-term success of businesses, their ability to grow and innovate, must be attributable to more than luck or circumstance.[27] Successful businesses incorporate systematic learning processes into their long-range strategic planning.

To date, however, this research question has not been adequately explored in the literature on organizational theory as it pertains to public institutions.[28] Hall argues that organizational learning is a deliberate attempt by governments to make better decisions by combining the lessons of past policy with new information and situations.[29] Richard Rose argues that policy learning and change are products of lesson-drawing whereby institutions (such as a legislature) learn from the policy experiences of others.[30] Ray Rist insists that "organizational learning can be shown to exist in the public sector," although case studies of successful organizational learning in public organizations are limited to studies of well-defined bureaucratic organizations (such as the post office) but not to more diverse, less tightly coupled organizations such as schools or legislatures.[31]

This is problematic given the policy logic inherent in educational reform, that reform is improvement. For example, Michael Fullan, a leading expert on school change, asserts that "educational change is a learning experience."[32] Richard Elmore and Milbrey McLaughlin suggest that reforms fail because policymakers do not take into account the lessons of past reforms.[33] If public organizations can and do learn, then possibly education reform could produce its intended objective, the improvement of practice, if internal organizational learning processes are incorporated into the institutional design framework.

Not surprisingly, proponents of organizational learning have come under intense criticism from scholars working in a variety of disciplines. Critics argue that models premised on organizational learning rely too heavily on an idealized or rationalized policy paradigm that fails to adequately capture the uncertainty and interpretive nature of policymaking, exemplified by garbage can models of the policymaking process.[34] Paul Sabatier and Hank Jenkins-Smith argue that theories of organizational learning are overly dependent on complete information and on the existence of a stable subsystem of policy actors.[35] However, policymaking often takes place under conditions of uncertainty.[36]

Furthermore, near-constant turnover in the political system makes organizational learning problematic. Political organizations such as state legislatures regularly experience turnover. While much has been written about the success of incumbents in retaining office, legislative turnover remains a challenge to organizational learning. For example, 14 states experienced turnover rates in state legislatures of between 20–29 percent from 1994–1996; another 27 states experienced turnover of less than 20 percent during that same period.[37] On the other hand, 9 states had a turnover rate of 30 percent or more. Even a low turnover rate of 15 percent, however, means that a third of a state legislature will be replaced every few years.[38] In 2003, nearly one-third of the members of the Texas House of Representatives and Senate will be freshmen representatives. Under these conditions, organizational learning is difficult when the players are constantly changing. Few state legislatures are able to maintain an effective institutional memory, particularly when representatives are bound by political promises. Technical rationality—crafting the best policies—is not always the top priority of politicians, although models of organizational learning assume otherwise.

Leeuw and Sonnichsen question the rationality upon which organizational learning is premised: that "the relation of information to action was presumed to be linear and straightforward."[39] The authors assert that "evidence now suggests that this linear assumption is naive" because "the incorporation of information into the knowledge base of an individual or an organization is selective, sporadic, and temporal."[40] Christopher Bosso agrees, noting that the policymaking process is messy, far less linear, and more complicated that current models suggest.[41] Decisionmaking in organizations, especially public organizations, "takes place in a context where 'rational information' is but one among many contending forces."[42] These forces include conflict over the values, goals, and objectives of the organization.[43]

Resolving questions over conflicting values "involves a contest of personal preferences in which reason plays little role, political criteria rather than scientific principles govern the determination of value disputes."[44] Postmodern scholars have questioned the very idea of whether "there is a real, knowable past, a record of evolutionary progress of human ideas, institutions, or actions."[45] Organizational learning assumes that there is a real, knowable past and that progress based on rationality is possible, indeed likely. While this learning may hold true in some instances for private organizations, it may be less valid when applied to public organizations. As Mary Hawkesworth argues, "the central questions of politics do not conform to an illusory model of knowledge in which there is one and only one correct position supported by an uncompromising logic derived from the necessity of truth."[46] Crafting education policy is as much a political process as it is a technical enterprise—posing even greater challenges for policymakers seeking to reform schools and learn from past practices.

Ultimately, public policy rests upon choices made by the political community and not on purely rational or technical grounds. Political conflict revolves around the competing values of participants. As Larry Cuban observes, "solutions are supposed to end problems; however, conflicts over values produce compromises that are struck and restruck over time."[47] This conflict strikes at the heart of policy change vis-à-vis organizational learning. Donald Warren argues that educational reform lacks institutional memory[48] and, as such, current reforms like the movement toward charter schools and vouchers may not build upon the lessons of previous experience. Policy change may be viewed as a series of tactical moves designed to ease political tensions and mediate conflict over competing values, rather than as the logical product of rational decisionmaking.[49]

Therefore, policymaking is far less predictable than organizational learning theorists suggest.[50] Even Rist acknowledges the difficulty of applying this concept to explain policy change in government.[51] Within government, decisions seldom stay made "because governance is the constant process of balancing demands, needs, and political pressures, decisions are most often temporary and partial, not permanent and complete."[52] This conclusion is supported by Gary Mucciaroni who observes that "the evolution of a major public policy . . . is inelegant, complex, and changeable."[53]

Organizational learning is much easier in highly technical areas such as airline deregulation, communications, and energy policy, where science more clearly identifies best practices, than in areas such as education, where substantial disagreement exists about virtually everything—from pedagogy and curriculum to school finance and governance.[54] Consistent with properties of schools as loosely coupled organizations, goals are multiple, often conflicting, and indeterminate. Little agreement exists on best practices—what is the "one best system" to educate children?[55] Finally, the participation and involvement of organizational members is fluid (the number of participants is exceeded only by the various degrees of involvement of stakeholders). Organizational learning under these conditions is exceedingly difficult.

Despite these critiques, Rist insists that decisionmaking in the public sector is more than constant improvisation and ad hoc decisionmaking.[56] The point "is not that learning is not taking place, but rather that it is taking place in more subtle and differentiated ways."[57] Unfortunately, "many of the fundamental elements of such learning remain conceptually unclear and, as a result, the entire phenomenon of experience-induced policy change remains difficult to operationalize."[58] Colin Bennett and Michael Howlett assert that

> methodologically, one of the major problems involves finding solid empirical work that unequivocally demonstrates that *X* would not have happened had "learning" not taken place. The conceptualization of learning as a kind of intervening variable between the agency (independent variable) and the change (dependent variable), however, may never be successfully operationalized . . . We may only know that learning is taking place because policy change is taking place.[59]

This leaves it up to the research community to evaluate the effects of policy, to determine whether policy change leads to better policies and better schools.[60] Unfortunately, reviewing the history of educational research, the research community rarely reaches consensus about what education policies work best. For example, researchers have spent more than three decades evaluating Head Start, yet they are still unsure whether the program works or whether it is cost effective. Similar controversies abound in other politically contentious education policies as well, including efforts to reduce class size, teacher certification, and bilingual education.

Critiques of organizational learning suggest that organizational learning and policy change are inseparable. If so, it is unclear whether all policy change is evidence of organizational learning. The theory of organizational learning is meant to explain policy change. The result, however, is that policy change leads us to explain organizational learning. In effect, the event illustrates the theory, but the theory does not explain the event. In light of these criticisms, can the theory of organizational learning be used to explain the spread of charter schools and the failure of voucher plans to pass legislative muster?[61] If so, then generalizations may be made to the educational reform movement *writ large*. If not, then other theoretical explanations such as institutional theory or advocacy coalition models may explain more accurately the movement for school choice.

ORGANIZATIONAL LEARNING PROCESSES IN STATE POLITICS

In her overview of charter schooling in the United States and Canada, Sandra Vergari observed that "research does not typically occupy a privileged position in the policy-making arena."[62] Concerned with the lack of empirical data on the effectiveness of school choice, particularly given the movement's growing popularity, the National Research Council (NRC) recommended a multidistrict, ten-year study of school vouchers. Edmund Gordon, a member of the NRC committee that wrote the report, expressed concern that as Congress, state, and local governments explore various voucher plans, there is insufficient empirical evidence available upon which to make informed decisions.[63]

Although use of academic research to inform policy is rare, particularly in "soft" social science areas such as education, research is sometimes used to guide state policymakers in crafting better education policies. For example, Tennessee lawmakers used the value-added assessment system developed by William Sanders as the centerpiece of the state's school improvement efforts.[64] Several states, most notably Florida, Kentucky, South Carolina, and Texas, have undertaken comprehensive systemic reform initiatives and improved their educational systems over time, based in part on policy feedback.[65] Educational reformers at the federal, state, and local levels are "working to create coherent policy systems by aligning key policies to support demanding learning goals."[66] Incoming reports on these initiatives suggest some degree of organizational learning is occurring, although the research community is not unanimous in its endorsement of these reforms.[67]

Recent state systemic reform initiatives are based on the development of highly complex, technical accountability systems, reflecting an attempt to more tightly couple what has heretofore been a loosely coupled system.[68] Thus, we might hypothesize that organizational learning in education is more likely to occur in highly specialized, technical areas such as educational accountability and less likely to occur in highly politicized, ambiguous areas such as school choice.[69] This hypothesis is similar to two hypotheses of policy learning developed by Hank Jenkins-Smith and Paul Sabatier who assert that (1) "Problems for which accepted quantitative data and theory exist are more conducive to policy-oriented learning than those in which data and theory are generally qualitative, quite subjective, or altogether lacking" and (2) "Problems involving natural systems are more conducive to policy-oriented learning than those involving purely social or political systems because in the former many of the critical variables are not themselves active strategists and controlled experimentation is more feasible."[70] Controlled experimentation in education, particularly in the areas of curriculum and governance, is rarely possible. Thus, one would not expect that organizational learning processes would explain the movement toward school choice, although none of the hypotheses offered above rule it out.

Research Utilization on Charter Schools

Let us now examine the extent of research utilization and consideration of past policy performance with respect to the decision to create charter schools in Texas and other states. Was the adoption of charter schools the product of organizational learning by state legislative leaders? Was research and evaluation used, and if so, to what extent and in what way? If not, then is organizational learning merely a fashionable theory, inapplicable in the rough and tumble world of power politics? What are the implications for educational reform and change?

With respect to the creation of charter schools in Texas, a former senior staffer with the Senate Education Committee stated that

> research did play a role. I don't think it would be the kind of research that we would understand, you know, where it would be hard academic research where you would have variables and where you would determine cause and effect and that sort of thing. I think it was research like "What are other states doing?" Anecdotally, "What do we understand has been good?" as it relates to what they're doing and that sort of thing. So I think there was some attempt to look at some studies and reports and that sort of thing as it relates to the experience of other

> states and I think there was an attempt by the Joint Select Committee[71] to actually invite experts from out of state to come in and share their experiences through testimony with the committee and any supporting documentation they had, but there was no academic research that we looked to guide decisions. I think more than anything else, what we're looking at is, "Let's provide our own laboratory by which to make informed judgments over the success or failure of our own experiment."

Another senior staffer noted that Senators Teel Bivins and Bill Ratliff, both Republicans, read everything with respect to research reports. The staffer stated that Senator Bivins believes "if anything should be subject to the scientific process, the educational system should be." But she noted, "the problem with the research is you've got people on both sides writing lots of articles in scholarly magazines and so I think it depends on your bias." She observed that it would be more helpful if the research community was more unified in its findings. She also noted that the preliminary findings of much research is of little use to policymakers operating under severe time constraints.

When asked where the idea for charter schools came from, she stated "probably they had seen it in other states." This assertion is consistent with another staff member's observation that "there, I think, had been other states that had experimented with charter schools and so it had been a concept that legislators at the time were increasingly becoming familiar with. It was a fairly new concept. For example, Arizona and Michigan are two states that I recall which had a limited charter program. So I think that kinda was where the idea came from, from sort of modeling after what other states had done." A member of one of the state's four major teacher organizations concurred, observing that "there is a charter movement going on across the nation." He believed the idea for charter schools in Texas came as a result of the national movement.

As the statements above suggest, little consideration was given to research on the effectiveness of charter schools or of the past policy performance of charter schools in other states, as least as research is commonly understood. Only a vague recognition occurred that other states were experimenting with charter schools and some policymakers expressed a desire to experiment with them in Texas as well. Surprisingly, the data suggested little debate over whether charter schools should be created, only debate concerning the expansion in the number of charters granted and the form those charters would take. Review of legislative testimony from the 1993 and 1995 legislative sessions revealed little dispute about the creation of charter schools themselves, suggesting that organizational learning had little to do with the adoption of this educational reform. This finding is particularly striking given evidence that performance on state exams had been steadily improving during the period of legislative debates over the creation of charter schools in Texas.[72] Drawing from data provided by the Texas Education Agency, student performance on the TAAS increased in grades 3–8 from 2–9 percent per grade level during the 1995 and 1996 academic years.[73] Why would policymakers seek to reform a system that was showing evidence of improvement? Models of organizational learning provide no answers to this question. By ignoring such evidence and structuring or framing the debate in a negative "school failure" fashion, the question became not whether public education should be reformed, but rather how.

Shortly after the Texas legislature passed a charter school bill in 1995, efforts were made to strengthen the law, expanding the number of charters that could be granted by lifting the cap. However, supporters of public schools, including various teacher groups, urged caution and said expansion of charter schools should be delayed until their performance could be adequately evaluated and compared with public schools.[74] John Cole, president of the Texas Federation of Teachers (TFT), stated, "They [charter schools] haven't taught one kid yet, so we don't know if this is a successful endeavor or not. We must always bear in mind that we are experimenting with other people's children."[75] In the minority report issued by Henry Jennings of the Joint Select Committee, Jennings insisted that the law limit the number of charters granted to allow "a better opportunity for field testing the idea before wholesale expansion promotes a practice that may not be educationally sound."[76] Doug Rogers, executive director of the Association of Texas Professional Educators (ATPE), stated "we would not like for him [Governor Bush] to expand the charter school program until we've had time to see how they have performed."[77] John O'Sullivan of the Texas Federation of Teachers argued that there has been little time to evaluate the effectiveness of charter schools.[78]

During the last four legislative sessions, controversy raged over the quality and viability of charter schools.[79] Senator Ratliff stated, "I am very pleased at the nature of the charters that we've had."[80] A spokesperson with a major pro-voucher group that also supports charter schools said, "It was very very difficult to bash the charter school movement when you looked at the characteristics and the demographics of the students who are attending the charter schools that do exist in Texas. What you find is that 78 percent of those students are at-risk children. Most of them [charter schools] are drop-out recovery programs and they're performing a service to kids who are most needy."

In an interview, an official with the Texas Elementary Principals and Supervisors Association (TEPSA)[81] stated "charter schools sound fine but there is no evidence anywhere that they have been successful in increasing academic achievement. We heard a lot of anecdotal evidence in the testimony. The bottom line of our testimony was until there is some hard evidence, we think that a major expansion is a mistake. We suggested that they go up by twenty until they have some hard evidence." In another interview, a senior official with the Texas Federation of Teachers said his organization opposed the expansion of charter schools because of a lack of evidence that charter schools were effective, noting that in a committee hearing state Representative Paul Sadler commented that they had just opened their doors.

In the original charter school legislation passed in Texas in 1995, up to 20 charter schools were authorized under S.B. 1 and charter schools were "sold" as a pilot program. The legislature authorized that an independent evaluation be conducted to determine the viability of the program—whether it should be subsequently expanded or disbanded. The debate over the expansion of the number of charter schools occurred during the 1997 legislative session. It is the nature of this debate that sheds the most light on the role, if any, of organizational learning in legislative decisions on education reform.

In an interview, a senior staffer for a Hispanic state representative noted that the original charter proposal was for a pilot study of 20 voucher schools to "test and determine whether it is worthy or not." A member of one of the teacher organizations stated, "there was concern about not having, like you know, open market on charter schools but rather having the limited number as it started out with so that we can learn from those and determine whether or not it's a program that should be continued or expanded on a gradual basis. There really hasn't been any time to [evaluate charter schools]. I don't see them being able to justify it [expansion]."

However, not long afterward (in the 1997 legislative session), the evaluation component was discarded and efforts were made to take the limit (cap) off the number of charters that could be granted. In an interview, a senior Republican senator noted that S.B. 1 provided for an independent review of the performance of charter schools but when the legislature went back into session,

> they (charter schools) just barely had their doors open so any research at that point wouldn't have made any difference. I think the reason we were able to expand it is that these were not white flight, that the vast majority of these are being created to address bilingual or at-risk or kids from bad socioeconomic circumstances. And the people who were originally afraid of them began to understand that this is, this may be a ticket out for some of those kids. So, I think it was more the experience of what type of programs they are offering as opposed to the results or the quality of them.

In an interview, a representative of another teacher organization said in her testimony before the Senate Education Committee, "OK, let's hold the line there [at 20 pilot charter schools]. We should hold the line until we have some data suggesting that they are working or not working." A senior official at the Texas Education Agency stated, "they [the Texas Association of School Boards] testified against the increase in the number of charters" as did all the teacher organizations. She said, "mostly what they [TASB] were saying is, 'Let's don't give any more [charters] until we are sure these are working.' "

However, contrary to what might be expected under conditions of organizational learning, the legislature voted to expand the number of charter schools from 20 to 100 without examining whether the reform was working, without waiting until any preliminary performance data were available.[82] This led a member of the Coalition of Public Schools (CPS) to state "we weren't pleased that this past year they expanded the number of charter schools. It was pretty stupid to go ahead and expand the number before the results were even in on the first twenty. That was, that was just not really good public policy."

In 2000, a panel of state lawmakers recommended a moratorium on new charter schools, citing poor student performance, unexpected closures, and financial troubles at some schools. In a report released by the House Public Education Committee, the committee said the State Board of Education did not adequately screen applicants and lacked enough workers to monitor the schools. The House passed a moratorium on the creation of new charter schools (capping the number at 215) and included additional regulations on established charters (H.B. 6). The Texas Public Policy Foundation, an organization that actively supports charter schools and vouchers, vigorously asserted that onerous regulations accompanying the bill would severely compromise the state's charter school law. However, the governor allowed the bill to become law without his signature.[83] Preliminary evaluations of charter school performance in Texas suggests that performance of charter schools is, at best, mixed, with student performance in charter schools lagging behind that of traditional public schools, although parental and student satisfaction in charter schools remains high.[84]

In most states, student performance in charter schools, compared to student performance in traditional public schools, is mixed.[85] As of October 2002, 194 of 2,874 schools (about 7 percent) ever given a charter have closed.[86] Often, however, the political response is to amend the original charter school legislation, expanding the number of charter schools in a state. In Texas, "in response to mixed reviews of the academic and financial performance of charter schools, the SBE [State Board of Education] recommended that the legislature grant no additional charters until the existing charter schools had proven successful."[87] Reports of the poor performance of students enrolled in charter schools have "fueled critics' claims that the state has rushed too quickly in expanding charter schools beyond the twenty schools originally authorized under S.B. 1."[88] However, a year later, "under political pressure, the SBE announced the creation of three new charter award cycles. A December 2000 House Committee on Public Education report criticized the SBE for granting some charters under political pressure from the Republican leadership."[89] In response to criticisms of the poor performance of charter schools in Texas, Teel Bivins, chair of the Senate Education Committee, stated, "This is a program that doesn't lend itself to instant evaluation."[90]

When the Ohio legislature passed charter school legislation in 1997, the law mandated five evaluations of the program over five years.[91] Based on preliminary results of student performance in charter schools, after two years of operation, passage rates of students in charter schools were well below state averages for traditional public schools.[92] In 2000, only 5 percent of charter school students who took the fourth-grade proficiency test passed all five sections, compared to 31 percent passing in traditional public schools; only 3 percent passed all sections of the sixth-grade test, compared to 35 percent for traditional public schools.[93] In response to the report, Tom Mooney, president of the Ohio Federation of Teachers, remarked, "If it was the public school's fault, you'd expect to see the kid soar like an academic eagle after getting to the charter school. We're not seeing that."[94] However, Clint Satow, director of the Ohio Community School Association, a charter school advocacy group, responded that, "They had the kids for six months and then they're tested. It's too early to see how these schools are doing until we see how they perform over a period of two, three, four, or five years."[95]

Satow accurately captured the tone in many state legislatures when he stated, "The debate in the legislature now seems to be: Do we slow down and study things, or do we go forward until we have a credible reason to think otherwise?"[96] While many legislatures, including Congress, employ the rhetoric of organizational learning and routinely create pilot or test programs, the nature of the political process mitigates against thoughtful, deliberate action. Once created, programs tend to become institutionalized, codified into state law, and expanded. Rarely are pilot programs carefully evaluated, and even when they are, politicians frequently ignore the results. Even programs that are wildly ineffective are rarely discontinued through the political process.

In every state, modifications have been made to charter school laws. Sometimes, these changes are the product of political conflict and pressure. Groups that were unable to block or prevent the creation of charter schools seek to restrict their expansion or impose limits on the scope of the initial charter legislation. In other cases, changes reflect lessons learned in the implementation process—itself a form of organizational learning. This highlights the multiple forms or types of organizational learning possible in political systems. At least two types of organizational learning are possible. First, organizations may learn which programs and policies produce optimal outcomes—what works best and why. For example, which school reform models are most effective? Should policymakers implement bilingual education or English-immersion policies? Programmatic questions such as these are one form of learning. The second form of organizational learning occurs during the implementation and evaluation phase of policymaking, as policymakers seek to tweak or modify existing policies in order to improve program effectiveness. Since laws are often vague, problems come to light in the implementation phase that are subsequently addressed during program modification or revision. The following examples highlight this second form of organizational learning.

California's charter school law has been amended several times in the past decade. The law "has evolved in response to lessons learned from the initial implementation process and also from the feedback of multiple stakeholders."[97] Several states have made revisions to the charter school application process.[98] In Texas, revisions include full financial disclosure and criminal background checks on all employees.[99] High-profile abuses in Ohio's 1997 charter school law led to substantial revisions of the state's charter legislation. On other occasions, changes reflect political battles waged among interest groups and political coalitions—groups that lost the original battle opposing charter schools seek to reassert their interests through amendments to the original charter legislation as well as through the implementation process. From a political conflict perspective, the game of education politics never ends since organized interests are ever present.[100]

Research Utilization on Vouchers

Let us now examine the extent to which state legislatures utilize research on vouchers in their deliberations. Using Texas as an illustrative case study, was the legislature's failure to adopt voucher plans in each of the last five legislative sessions the product of organizational learning, whereby policymakers evaluated the performance of past policies and practices and rejected proposed reforms based upon the experiences of others?

Unlike the debate over the creation of charter schools in Texas, the debate over various voucher plans included many references to research and policy studies. A study of choice programs in 20 states by the Texas State Teachers Association concluded that school choice was not the solution to the problems of public education.[101] Richard Kouri, then president of TSTA, stated "there is not one single school district in all these twenty states where student achievement has risen as a result of 'choice' alone."[102] TSTA found that few parents actually exercise their choice options. Accordingly, TSTA rejected vouchers and endorsed only one form of school choice—charter schools.[103] Kouri stated, "In all of our findings, only charter schools offer a prospect for true educational reform."[104]

Reflecting on the "evaluation" of the pilot program for charter schools, a member of a teacher organization stated in an interview that "there is concern among voucher opponents about a pilot voucher program being expanded before being properly evaluated [like what occurred with charter schools]." Nancy McClaran, executive director of the Texas Association for Supervision and Curriculum Development (TASCD), cited research by the Economic Policy Institute[105] that found that "parents find it very difficult to choose schools based upon academic quality."[106] In an editorial, she suggests that "choice" may well be an illusion for many parents.

Several voucher proponents in the legislature said that research does play a role in legislative decisionmaking. In an interview, a Republican House member stated, "I think it [research] plays a huge role. I look at it really carefully. It has a huge impact on what kind of bill we are going to write. We are looking real closely at what goes on in Cleveland and Milwaukee and elsewhere to see if there are any pitfalls we need to be aware of, if there is any way we can improve the system to make sure that what we do pass encourages the public schools to become better and doesn't harm them." Arguing in support of a pilot voucher program, Senator Bivins stated "my goal . . . is only to provide a forum to see which side is right. I am not the standard-bearer for the voucher group, nor for those opposed. All I want to do is provide the mechanism to see if it does indeed advance student learning or harms student learning."[107]

Bivins's comments echo earlier statements made by Governor Bush after a 1995 voucher plan failed to get through the legislature. Bush stated, "I still am very strongly in favor of the concept of the state trying it [vouchers] out and determining whether or not this interesting concept would work."[108] Bush favored a pilot voucher program specifically targeted at children from low-income, disadvantaged families that would allow public monies to be used to pay tuition at private schools.[109] In a press release, state Senator Jeff Wentworth stated, "I'm not sure who is right [whether vouchers work], but I do believe a limited-time, pilot program in a few counties to try it out to see whether it might work is an idea worth pursuing. If it works, we can expand it. If it doesn't, we can stop it. But we'll never know for sure until we try it."[110]

In public testimony, Ed Adams, a member of Texans for Education, a coalition of major Texas employers working to improve Texas public schools through sound public policy, said he believes in experimentation (such as pilot programs for education reform) but wants to see the results of these initiatives before committing to new reforms. He stated, "it would bring it [debate over vouchers] to one conclusion or the other. It may prove that it does improve school performance and that we might want to recommend doing more." When questioned in an interview about the role of research in the debate over vouchers, a senior staff member of the Senate Education Committee replied, "we won't know if a voucher program will work unless we try a limited pilot program."

Taken at face value, such comments suggest at least the possibility that organizational learning may in the long run influence the adoption or non-adoption (as well as expansion or contraction) of voucher plans. Certainly, the political rhetoric suggests the possibility for organizational learning to occur. The strongest evidence against such a scenario is the Texas legislature's handling of the charter school reform, where the legislature created an evaluative component and subsequently chose to expand the program before evaluating its effectiveness. An evaluative component was included in a voucher bill introduced by Senator Bivins during the 1997 legislative session.[111] CSSB 1206 would have authorized the State Board of Education to designate an impartial organization to annually evaluate the pilot voucher program using criteria such as test scores, grades, attendance, disciplinary incidents, socioeconomic data, and parental satisfaction.[112] However, given legislators' failure to consider the evaluative component pertaining to charter schools included in S.B. 1, it is entirely possible, some would say likely, that the legislature will act in a similar manner when acting on any evaluative component included in a pilot voucher plan.

If so, then the question is raised whether research matters at all in legislative decisionmaking, or if it simply is used as policy argument. In general, members of the Texas legislature believe that policy research has a real impact on legislative decisionmaking, as the testimonials above suggest. However, many members of various interest groups (both for and against various school choice initiatives) question this "objectiveness" and believe that research was used in the debate over charters and vouchers as policy argument. In an interview, an official at TEPSA questioned the role of research in policymaking. She believes the legislature makes decisions without looking at research and

> I think that's very true and I think the other thing they do is they use research to make their point. For example, this whole voucher thing at the last session, they kept throwing out how many low-performing schools there were. Well, to get on the low-performing school list, there are a lot of ways you can get on there. It doesn't mean it's a bad school. Not necessarily. But they throw out these numbers and they appear to be big and inflated. [She mentioned that the governor's daughters attended one of the schools listed as low-performing.] Why? One subset. Hispanic students exceeded the dropout rate by two kids. In the charters thing, they didn't pay any attention to any of the research that we gave them. They didn't wait to find out the results of the TAAS [Texas Assessment of Academic Skills].

The head of a major anti-voucher coalition stated in an interview:

> In terms of research, we did pay for a poll. We came up with numbers that two-thirds of the people in Texas did not like the idea of taking tax

> money for vouchers. And we shared this at the Capitol and it helped to turn some legislators around. Plus, it showed that people of both parties oppose vouchers. Poor people oppose vouchers. That was one way that I think that the research helped some people. And particularly in the House, I think we had four Republicans who voted with us. And I can think of one of those particularly who used this information as part of his justification.

In this instance, however, the "research" does not take the form of whether the reform is effective, but rather functions as a type of opinion poll about what constituents favor or oppose, the results of which are used for purposes of policy argument to justify political positions.

Consistent with this viewpoint, a member of a pro-voucher interest group said, "You don't convince through research. It's very rare that you would convince them [members of the legislature] to completely change their worldview on, you know, competition. What research does is that it gives them some proof that there is a problem." This is consistent with the statement made by a House member in an interview who, when asked if she would change her position on vouchers if the research conclusively demonstrated that voucher plans are ineffective, replied, "I guess it would depend on what the reason was for it [vouchers] failing. It's kind of hard to image a reason why choice would fail."

Discussing the role of research in policymaking, the head of an interest group stated, "it was pretty appalling to me, you know, that people would vote for things that they didn't even really understand. I mean, we're talking about not looking at research, we're talking about [not] reading the bill. It's frightening." The spokesperson for another interest group had a different perspective on the impact of research on legislative decisionmaking. She noted that:

> Normally, a small special interest goes and convinces, either at the local caucus or whatever, that this is something that we want as part of our plank, and then it grows and grows and grows. But, they [members of the legislature] don't always do their homework because there is one thing I have found and they can't. We sit back here and we're thinking public education and that's it. They got to think about everything: highways and roads and health, etc., everything. Sometimes sitting in those committee meetings, you can always tell that a lot of times what is rudimentary to us can be eye-opening for committee members because they've got so many things to deal with that they have not, they and their staffs have not necessarily had the time to do research and do all the background. Sometimes, they have not done their homework.

When listening to state policymakers talk about school choice, they clearly use the language of organizational learning—making frequent use of lingo such as "experimentation" and "evaluation." For example, Scott Milburn, a spokesperson for Ohio Governor Robert Taft, said Governor Taft "thinks we need to continue the experiment [the Cleveland voucher program] and evaluate it."[113] Former Governor George Voinovich, an ardent supporter of vouchers, said of the Cleveland voucher program, "Let's give it two or three years and see if it works."[114] State Senator Robert Gardner, an Ohio Republican who chairs the Senate education committee, stated that, "I know parents are happy with it [the Cleveland voucher program], but I haven't seen much accountability data. We fund a lot of these pilot projects, but the problem is we don't take the necessary steps to validate the results."[115] Gardner's comments are typical of how legislators craft policy in the political arena.

A major difficulty with viewing policy change through the lens of organizational learning is that the framework assumes that policy is crafted absent politics—that each policy represents the best thinking of the legislature or is the optimal product of the policymaking process. On the contrary, policy is the product of extensive political bargaining, negotiation, and compromise. The experience and statements of Pennsylvania Governor Tom Ridge offer a perfect exemplar of this phenomenon. Governor Ridge floated two different voucher proposals before the state legislature, trying to determine which plan, or combination of plans, would muster the majority support necessary for passage. Ridge stated, "What we need to do . . . is come up with the right combination that solidifies a majority of votes in both chambers."[116] Ridge continued, "It's a negotiating process we've been involved in to get the requisite number of votes."[117]

After both proposals failed to garner enough votes, Governor Ridge put forth a bizarre compromise voucher plan permitting each of the state's 501 school districts to establish its own tuition voucher program. However, districts would not be required to adopt or participate in the voucher program and no state funding would be forthcoming; each district would have to pay for vouchers from its own operating budget. Furthermore, students in only one district, Chester-Upland (which was taken over by the state and is governed by a state-appointed control board), would be immediately eligible to receive vouchers. Districts identified as "academically distressed" would be given two years to improve; failure to demonstrate improvement after two years would make students in those districts eligible to use vouchers.[118] Ultimately, this compromise pleased no one because it fell far short of what voucher proponents advocated, and failed to win the support of Democrats opposed to vouchers in principle.

In their rush to maintain or expand school choice initiatives, many state legislatures ignore evaluation data or remove evaluative criteria from initial pilot programs. For example, in 1995, when the Wisconsin legislature expanded the Milwaukee Parental Choice Program, "it effectively removed the requirement that the program be evaluated."[119] In 2000, amidst reports of concern over inadequate oversight of school choice programs in Ohio, an audit of Milwaukee's voucher program was conducted by Wisconsin's Legislative Audit Bureau.[120] The results provided ammunition for both supporters and opponents of the voucher program. While the audit revealed that the program enrolled students from low-income families (as intended), it faulted the state for its failure to collect student-performance data on students enrolled in the program.[121] In 1995, when Milwaukee's voucher program was expanded to include religious schools, state legislators removed the requirement that students receiving vouchers must take the same state-mandated tests as public school students—making it much more difficult for the state to evaluate the impact of the program on student achievement.[122]

ORGANIZATIONAL LEARNING, SCHOOL CHOICE, AND THE POLITICAL DYNAMICS OF POLICYMAKING

The political history of school choice in the 1990s does not appear to support the notion that choice is the product of organizational learning, as work by Baldersheim, Hall, and Rose might suggest.[123] Although an evaluation of the pilot charter school program was mandated in S.B. 1, the Texas legislature ignored the independent review board, headquartered at the University of North Texas, charged with evaluating the effectiveness of the policy change. Had the legislature deferred action until the preliminary results of the review were complete, then perhaps organizational learning might have occurred. However, since the legislature moved so quickly to expand the number of charters without consideration of the effectiveness of the reform, coupled with little or no systematic effort to evaluate the reform in other states, suggests that the adoption of charter schools was not the product of organizational learning, whereby policymakers evaluate the performance of past policies and practices and initiate reforms to improve upon past practice. Political pressures appear to have won out, with the addition of three new award cycles.

Policymakers are under severe organizational constraints that may limit the extent to which organizational learning processes may be incorporated into legislative decisionmaking. Unlike other public organizations, such as schools, where policymakers are selected based (presumably) on ability or merit, state legislatures are composed of politicians elected on the basis of popularity.[124] While many highly competent individuals serve in state legislatures throughout the country, the demands of politics are seldom conducive to rational, value-free decisionmaking. The political system, with its built-in bias for compromise, often produces not the best policy (in terms of effectiveness) but rather the policy that can garner enough votes to get through the legislative maze. The complicated political calculus necessary in crafting and passing legislation is largely absent from models of organizational learning, which assumes a degree of rationality and objectiveness often absent in the political system.

These findings question, although they do not preclude, the likelihood that organizational learning processes are at work in the debates over charter schools and vouchers. In the debate over vouchers, research appears to have been used primarily as policy advocacy. Legislative testimony, interview data, and other documentary sources suggest that although state legislatures appear to have put in place the conditions for organizational learning to occur (via evaluation), the lack of attention given these processes and subsequent efforts to expand programs absent such evaluation leads to the conclusion that organizational learning cannot account for the adoption and expansion of charter schools, and the resistance in many states to vouchers.

The creation of an evaluation component for charter schools and a similar component in the proposed voucher bill clearly indicates that the Texas legislature was putting into place a structure within which organizational learning might occur. However, the legislature's actions with respect to the expansion of charter schools, coupled with its likely expansion of any voucher pilot prior to completing an evaluation of the program, leads to a conclusion made earlier by Hawkesworth who found that "political criteria rather than scientific principles govern the determination of values disputes."[125] This view is consistent with Cuban's observation that "conflicts over values produce compromises that are struck and restruck over time."[126] Despite Senator Bivins's and then-Governor Bush's insistence that a pilot voucher program is necessary to determine whether the concept would work, scientific criteria are unlikely to play a major role in resolving the dispute. The effect of research seems to have been, as Callahan and Jennings found, "indirect—one small piece in a larger mosaic of politics, bargaining, and compromise."[127] Particularly in the case of vouchers, research seems to have taken the form of policy argument, a finding consistent with earlier studies of policymaking by Frank Lutz, Deborah Stone, and Carol Weiss.[128]

IMPEDIMENTS TO LEARNING FROM SCHOOL CHOICE

Many state charter school laws contain provisions for evaluation. However, evaluating whether charter schools improve student performance is a tricky undertaking. Although Minnesota created the nation's first charter schools in 1992, charter schools in several states have been in operation for less than five years. As a result, initially low student performance in charter schools may reflect the quality of the education students received in their previous public school.[129] Even today, the number of charter schools nationwide is small, approximately 2,700. Within each state, wide variability exists in how long each charter school has been in existence, making within-state comparisons difficult. Typically, a state's charter schools are compared against all the state's traditional public schools, producing a comparison of a small "n" (student performance in charter schools) with a much larger "N" (student performance in traditional public schools). Since most charter schools are located in urban areas, comparing them as a group with public schools statewide may be unfair.

Charter schools in several states such as Texas enroll higher percentages of at-risk students than are found in traditional public schools; in fact, many charter school laws give special preference for applicants serving at-risk children.[130] Coupled with the fact that charter schools are significantly smaller than traditional public schools, "many small charter schools target populations so specific that there is no traditional public school counterpart to use as a basis of comparison."[131] It takes only one or two low-performing students for a school to be classified as low-performing or failing. Some policymakers assert that "it takes at least five years to determine the effectiveness of a new charter school."[132]

Joseph Viteritti argues that existing choice programs cannot be evaluated and compared because "most have been designed to limit real competition." For example, many states impose limits on the number of charter schools allowed, and some provide less per pupil funding than for students enrolled in traditional public schools. Voucher programs contain similar caps, with per pupil funding (the amount of the voucher) less than the per pupil allocation for public school students. For example, in Cleveland, "per capita public spending for children who attend regular public schools is $7,746; for charter school students [in Ohio], it is $4,518; and for students who use vouchers, it is $2,250"—effectively creating an unfair competitive arrangement and imposing significant opportunity costs on students (and schools) that participate in school choice programs.[133] Such limitations make problematic valid comparisons and evaluation.

CONCLUSION

Viteritti trenchantly observed that, "Behind the school choice debate that has occupied policymakers so intensely for the past 10 years is the fantastic notion that some day a group of dispassionate experts will objectively reach a judgment" to determine whether school choice works.[134] Sandra Vergari agrees, observing that "even if the [charter school] reform were found to display serious deficiencies, significant political support might still be forthcoming because the reform embodies values cherished by power elites and/or because continued support of the reform is deemed to be politically advantageous."[135] The available evidence from Texas and other states suggests that organizational learning processes are in place but largely ignored in the activities of state legislatures as they grapple with school choice. Perhaps the inclusion of pilot programs and evaluative components in choice legislation represents keen political strategy rather than a desire to actually evaluate programs prior to expanding them. In fact, a case could be made that casting controversial programs as pilot studies subject to review and evaluation makes them more politically salable.

These findings are consistent with a large, diverse body of research that suggests that politics, rather than rational, evidence-based decisionmaking, drives the policymaking process.[136] However, these findings question much of the research on organizational learning, as exemplified in the research of Rist and others who assert that governments learn important lessons from each other. In the case of school choice, state legislatures tend to respond to pressure from interest groups or policy entrepreneurs, with less attention to issues of organizational learning. Even when state legislatures incorporate organizational learning processes into their policy activities, such as through the creation of pilot programs, demonstration projects, and evaluation components, they rarely base subsequent decisions on these evaluations—rhetoric notwithstanding.

In many respects, the tendency of state legislatures to ignore organizational learning about school choice is consistent with activity in other areas of state education reform as well. For example, David Cohen and Heather Hill conducted a detailed study of California's state-level effort to improve teaching and learning in mathematics. While the authors found that, contrary to earlier research, state-level education reform works—that it can positively influence teaching and teaching—the authors also found that policymakers "made no attempt to learn systematically about how the reforms played out in schools and classrooms."[137] Cohen and Hill go on to note that, "Neither the professional organizations that backed the California reforms nor the state education agency made any effort either to learn systematically from experience or to report findings to professionals, the public, and policymakers. Though professionals in these organizations were happy to help us, their approach to educational improvement did not seem to include systematic learning about the implementation and effects of their efforts."[138] Unfortunately, such reports are not atypical of state legislative decisionmaking.

Furthermore, even if state and federal policymakers wanted to incorporate organizational learning into their decisionmaking processes, they would be faced with a confounding mass of inconclusive, conflicting research. Even well-established programs such as Head Start, which has been around for decades and studied ad nauseum, fail to generate consistent research findings. In fact, given the inconclusive nature of virtually all education research, there is little agreement on what works best, where, and under what conditions, making it problematic for policymakers to learn. Learning is difficult, if not impossible, when the lessons themselves are conflicting and unclear. Finally, the highly politicized nature of state policymaking, with its systemic emphasis on bargaining, negotiation, and compromise, discourages organizational learning, often producing not the most effective policy but rather the "best possible policy" given political constraints, which is a lesson in itself.

Chapter 6

The Political Dynamics of School Choice: Lessons Learned

In his classic study of state politics and policymaking in the South, V. O. Key observed that, ultimately, politics boils down to a fundamental conflict between the haves and the have-nots—between disenfranchised groups seeking a greater share of the economic pie and others unwilling to part with what they have.[1] Political battles over school choice reflect this tension. While conservative ideologues lead the charge for vouchers and are joined by an assortment of urban minority (and predominately) liberal activists, together with many religious leaders seeking financial support for nonpublic schools, they face strong opposition from the traditional education establishment (teachers' unions, administrators, and school boards) and their (primarily) Democratic supporters in state legislatures throughout the country, coupled with some Republicans representing suburban and rural constituents.

An Idea Whose Time Has Come?

In their assessment of the shifting politics of school choice, Robert Bulman and David Kirp conclude that "in one form or other, school choice is here to stay."[2] During one of numerous debates on vouchers, Republican state Representative Kent Grusendorf concluded that "defenders of the status quo can run out their red herrings kicking and screaming all the way, but school choice is inevitable. It is just a matter of time."[3] A senior staffer to a Hispanic state representative noted that the Texas legislature has taken small, gradual steps toward vouchers. In an address to the Texas Parents and Teachers Association, Sandra Zelno, Pennsylvania PTA president agreed, stating that "some sort of choice is inevitable."[4] Even Senator Barrientos, a voucher opponent and charter school skeptic, admitted in a committee hearing that vouchers were inevitable sometime in the future.

Support for school choice, including vouchers, is no longer solely the province of right-wing ideologues, if it ever was. With an increasing number of liberal Democrats and people of color supporting choice, ultimately the traditional educational establishment and their allies in state legislatures must grapple with a movement that is spreading throughout the country, albeit slowly and unevenly. Ten years ago, school choice was not a major issue on the agendas of state policymakers. Today, charter schools have spread throughout most of the states, and voucher plans are under consideration in more than half the state legislatures. Scarcely a week goes by that school choice is not in the news.

Clearly, the supporters of school choice are a diverse lot and not always in agreement with one another about what type of school choice to support. Choice coalitions vary depending on the issue—charter schools, vouchers, tuition tax credits, etc. This finding is consistent with research by Hugh Morken and Jo Renee Formicola, who found "fragmentation, differences on what kind of school choice most advocates support, a lack of national leadership, and a need for more unity within advocacy ranks" in the school choice movement.[5]

Increasingly, school choice is linked to accountability reforms, as indicated by recent state and federal activity. At the federal level, under provisions contained in the No Child Left Behind Act, students in consistently low performing public schools can transfer to the public school of their choice. As part of revisions to Colorado's education accountability system, schools receiving a failing grade in two consecutive years will be converted into independent charter schools.[6]

As Michael Mintrom notes, "as evidence from many states now demonstrates, small steps like the introduction of open-enrollment plans have been followed by the introduction of other choice-based reforms, like charter schools. At every step these changes have developed constituencies prepared to go into political battle to keep them in place and extend their reach."[7] Mintrom views the rise of school choice as non-incremental policy change since initial, incremental steps—such as magnet schools, intra- and interdistrict choice, and charter schools—lead to more aggressive choice proposals such as school vouchers.

However, as Morken and Formicola observe, as a political movement school choice "is proceeding at a different pace in different places," with the outcome very much in doubt.[8] In several states, the momentum for expanding (or restricting) school choice ebbs and flows, responding to political currents and institutional dynamics. In Michigan, for example, after five years' progress in expanding choice options through open-enrollment laws and charter schools, the drive for further expansion has stalled. In 1999, the state legislature failed to lift the cap on the number of charters that could be granted by public universities. In addition, deep divisions within the state's Republican Party had brought momentum toward vouchers to a standstill.[9]

Given the recent U.S. Supreme Court ruling upholding the Cleveland voucher program, many observers believe that "a market for schooling is emerging and that the market's inexorable logic will have positive, transformative effects on public schooling."[10] However, Michael Mintrom and David Plank point out that, "While appealing, this view is faulty primarily because it discounts the ways that political dynamics so often serve to put the brakes on policy change."[11]

POLITICAL DYNAMICS AND POLICY LINKAGES

The coupling of various school choice initiatives, such as charter schools and vouchers, in state legislatures has significantly affected the political dynamics of the legislative process, facilitating the adoption and spread of charter schools. In only a decade, charter legislation has swept through three-fourths of the states, with additional legislation being considered each year. The rapid spread of the movement is attributable as much (if not more so) to the threat of vouchers than to the concept of charter schooling itself. As has been amply demonstrated in the text, charter schools generate substantially less conflict than voucher proposals precisely because groups opposed to expanded school choice devote nearly all their attention and limited resources to defeating vouchers.

For Democrats, charter schools offered a "convenient compromise" on choice.[12] As Morken and Formicola remind us, "compromise can look very attractive when the prospects of a bigger loss are real."[13] In California, for example, "the pressure placed on unions by voucher legislative proposals and by Proposition 226 [the initiative forbidding automatic deductions for union dues used in elections] helped to win support for charter schools as the lesser of two evils."[14] In many states, such as Texas and Pennsylvania, the threat of vouchers was so real that the traditional public education establishment, together with their allies in the legislature, was willing to make substantial concessions on charter school bills in order to block vouchers.

However, the strategy of using charter schools to take the steam out the voucher movement appears to have backfired. Instead, as many voucher proponents suggest, expanding school choice leads only to greater pressure for more choice. In Arizona, for example, after the legislature passed strong charter school legislation in place of vouchers, the pressure for vouchers continued unabated. Three years later, "voucher advocates were able to pass a tax-credit measure that encouraged taxpayers to support educational scholarship funds to help students attend private schools."[15] In most states, charter schools have not acted as a release valve, as a way to take the steam out of the voucher movement, and instead are paving the way for more radical choice initiatives such as vouchers. The "release valve" may have worked temporarily to forestall vouchers, but the momentum for expanded school choice continues unchecked. As a result, the next round of choice battles may be over how to limit the scope of voucher proposals under consideration in state legislatures.

Reviewing what we have learned about the political dynamics of school choice, we find evidence to suggest that models of political culture offer little guidance as to which states are most likely to adopt charter school or voucher legislation. While a generalized political culture serves to set the parameters of debates over choice, too many other contextual variables affect states' policymaking processes. In his study of charter school politics in Colorado, Georgia, Massachusetts, and Michigan, Bryan Hassel emphasized the "importance of the broader legislative context in which charter schools" are debated.[16] Similarly, models of organizational learning, with their myriad assumptions of rationality, evaluation, and control, fail to accurately describe the complexity of legislative policymaking.

Models drawn from institutional and advocacy coalition theory appear to have greater explanatory power. As the evidence presented in this book suggests, institutions play a major role in shaping the political dynamics of school choice. Far from being neutral, institutional bias clearly favors the adoption of less controversial proposals such as charter schools. Overcoming this institutional bias requires extraordinary commitment, resources, and, perhaps, a bit of luck, as external events (many unrelated to education at all) can cripple efforts to pass voucher legislation.

While the institutional context is critical, policymaking is fundamentally an interpersonal, relational process of bargaining, negotiation, and compromise. Effective policymaking requires building consensus, and coalitions are an essential component of that process. The evidence presented in this book demonstrates that broad-based coalition building is essential for the further expansion (or contraction) of school choice. As Mintrom and Plank aptly stated in their analysis of school choice politics in Michigan, "support for choice can no longer be accomplished on ideological grounds alone."[17] The work of choice lies in building sustainable political coalitions that will hold together over time. The shifting nature of many school choice coalitions discussed in this book suggests that advocacy coalitions may be more fluid and less stable than existing models of advocacy coalitions developed by Sabatier, Jenkins-Smith, and others suggest. However, the political dynamics of school choice clearly support the basic tenets of this theory of policymaking.[18]

ONGOING CONFLICTS: CHARTER SCHOOLS

As Joseph Viteritti observed, debate over charter schools has been transformed, fixed less "on the question of whether such laws should be enacted and more focused on the particulars of what these laws would entail."[19] Despite widespread bipartisan support for charter schools, significant differences exist among supporters as to what rules and regulations from which charter schools should be exempt. For example, the conservative Center for Education Reform and the nation's teachers' unions have a substantial difference of opinion on what constitutes good charter school legislation. Although both the National Education Association and the American Federation of Teachers have cautiously endorsed the charter school idea, with the NEA even providing financial support for some schools, both groups are firmly opposed to charter legislation that weakens teachers' collective bargaining rights.

Despite widespread adoption of charter school statutes in states throughout the nation, political battles over this reform continue unabated. As conservative critic Chester Finn has observed, a war is being fought over the expansion, even the continued existence, of charter schools. The future growth and expansion of the movement will depend on states liberalizing or strengthening their charter school laws—for example, by lifting caps on the number of charter schools that can operate in each state.[20] However, opposition is arising from several quarters, including unions, school boards (local as well as their state and national organizations), school districts, and professional administrator organizations—the traditional education establishment. For example, a recent issue of the United Federation of Teachers' weekly newsletter was chock full of anti-charter articles.[21] Lawsuits have been filed in several states. In Ohio, a coalition of local and national groups, led by the Ohio Federation of Teachers, has launched an all-out campaign to significantly revise the state's charter school legislation, even though the state's charter school law is not particularly strong.[22]

In many states, the traditional education establishment is "attempting to block efforts at expanding and strengthening charter school laws."[23] All states, except Arizona, which alone has a quarter of the nation's charter schools, set a limit on the number of charter schools allowed. Most states have fixed caps, although some states such as Texas have a "flexible" cap that allows unlimited charters above the state cap if the school is created specifically to serve at-risk children. When charter school legislation was reauthorized in Colorado, the cap on the number of at-risk charter schools was lifted and new regulations were adopted, making it easier to finance charter school facilities.[24] However, efforts "to permit entities other than school districts to authorize charter schools" have been consistently resisted and defeated by state lawmakers.[25]

As the charter school movement moves into its second decade, the politics surrounding charter schools is also undergoing change. With any policy, problems of implementation inevitably arise. Well-publicized abuses and problems with charter schools in several states have led legislators to modify and revise charter statutes, tightening up the laws, closing loopholes, and adding more restrictions and red tape (often under the auspices of making charter schools more accountable). In fact, the extent to which charter schools should be regulated is a major source of political conflict, as proponents believe charter schools are already over-regulated, while opponents view the schools as dangerously under-regulated.

As the political battles over charter schools continue, lawmakers in several states "have adopted significant amendments to charter school laws. To date, the amendments have typically been aimed at creating more fertile ground for the growth of charter schools."[26] A notable exception to this trend is in Michigan, where numerous amendments have made the charter school law more restrictive, including requirements that all teachers in Michigan charter schools be certified, no blanket waiver from compliance with the Michigan School Code, mandatory student assessment on state tests, and caps "placed upon the number of charters that can be issued by state universities."[27]

ONGOING CONFLICTS: VOUCHERS

As the events in this book demonstrate, when faced with an external threat such as vouchers, the traditional educational establishment will unite, forming coalitions to maximize their strength.[28] Despite the Supreme Court's ruling in the *Zelman* case, in some respects, the real battles over school vouchers are just beginning, as concerted attempts to pass voucher legislation are being made in state legislatures throughout the country. However, it would be a mistake to assume that states will rush to adopt voucher plans. Significant institutional and coalitional constraints remain. Coalitions supporting vouchers will need to be broad-based, including minority Democrats representing urban areas in which public support for choice is greatest. Second, voucher proposals will likely be equity-based pilot programs targeted at families from poor, under-served communities with persistently failing public schools.

According to Bulman and Kirp, "New equity-based voucher plans are most likely to emerge in urban school districts where black activists, conservative politicians, and foundations forge an alliance that contests the traditional liberal–labor coalition."[29] However, equity-based voucher plans risk "losing the support of existing private schools and middle-class parents, who may well find the new charter school regime a better bet."[30] Voucher plans must pose no threat to families in suburban school districts. Ultimately, even with changing constitutional standards, the fate of vouchers may turn on whether enough support can be generated from middle-class and upper middle-class families in suburban school districts.[31] State passage of pilot voucher plans targeted at low-income families depends on the willingness of middle- and upper-class suburban families to support expanding school choice to the poor and working classes. Suburban and rural state legislators in several states have noted little interest in their communities to support vouchers.

If middle- and upper-class families, who exercise choice by living in areas with excellent public schools or who already send their children to private and parochial schools, believe that vouchers will threaten state funding for suburban schools, change the complexion of those schools, or that voucher plans will bring unwanted state regulation to private schools, then voucher plans stand little chance of success. Since the vast majority of Democratic state legislators oppose vouchers, success of voucher initiatives depends on the support of suburban, Republican legislators—and that support is by no means assured. If suburban parents believe even for a minute that vouchers may potentially rob their public schools of funding, or that state intrusion may change the nature and character of private education, then voucher plans are doomed to failure. For it is not in the interest of middle- and upper-class suburban families to support any education reform that could potentially jeopardize the quality of education their children receive.

The momentum for vouchers, particularly plans targeted at low-income families in urban areas, is significantly affected by the condition of urban education. Parents and policymakers recognize that if urban schools did a better job of educating children, there would be little pressure for vouchers. Michael Fox, chair of the House Public Education Committee in Ohio, commented that the only reason why vouchers were tried in Cleveland was because the system was so insolvent, with student test scores abysmal for so long, state lawmakers were willing to try anything.[32] Accordingly, policy entrepreneurs and activists will play a key role in eventually passing vouchers. Although the traditional education establishment is effective at blocking legislation, the decades-long persistence of choice advocates (coupled with the substantial contributions made to key legislators) may eventually wear down the opposition enough to pass pilot voucher plans.

POLITICAL LEARNING: THE BENEFITS OF LOSING

What is remarkable about the school choice movement is how dedicated activists are to the cause, particularly vouchers. While charter schools were generally well received, and legislative efforts often met with quick success, the history of voucher battles is littered with repeated, sometimes devastating, failures. Vouchers have been resoundingly defeated in every statewide ballot initiative. Further, the election of staunch pro-voucher governors and more sympathetic legislators have been, as yet, insufficient to trigger a voucher revolution in the states. With the exception of a few legal victories, a couple of small pilot voucher programs in Cleveland and Milwaukee, and a small statewide initiative in Florida, voucher successes have been few. Yet, the movement continues, fueled by a combination of ideology, money, and (in the case of urban education) utter desperation.

However, in politics, benefits accrue even when a fight is lost. Choice proponents assert that despite repeated legislative defeats, the fact that voucher proposals were brought into the public debate constituted a political victory of sorts—laying the foundation for subsequent efforts.[33] Further, the political systems of the American states respond to persistent, applied pressure. Persistent effort by well-financed, single-issue lobbyists is rewarded in state policymaking.[34]

Some ideologues in the school choice movement have made serious strategic errors, such as adopting uncompromising "all or nothing" strategies. However, through active coalition-building voucher advocates are learning how to manipulate the political process more effectively to attain their desired objectives. For example, voucher advocates frustrated with recalcitrant state legislatures initially adopted a strategy of bringing about vouchers through popular initiatives and referendums. However, ballot initiatives and referendums have not proven an effective method of bringing expanded school choice via vouchers to the states. Because of their high visibility, voucher initiatives are more easily defeated by teachers' unions with large war chests and expensive attack ads.[35] Initiatives and referendums leave no room for the backroom political maneuvering, negotiation, and compromise so essential to passing controversial legislation.[36] Furthermore, adopting a political strategy used by civil rights leaders in the 1950s and 1960s, voucher organizations such as the Institute for Justice have turned to the courts and successfully tested the constitutionality of voucher plans—a decade-old battle that is expected to continue in several states.

Narrow ballot initiatives allow the traditional education establishment to join forces in a collective effort to defeat choice proposals. As a result, voucher advocates in many states have shifted their legislative efforts toward passage of comprehensive school reform, including expanded school choice as only one part of a multifaceted reform program. Such strategies stand a much better chance of success because such bills make unified, organized opposition more difficult. This political learning, what Sabatier and Jenkins-Smith refer to as policy learning, is an essential component of politics and is crucial to the success of school choice initiatives.

SCHOOL CHOICE AS ANTI-UNIONISM

Legislative battles over school choice highlight the weakening power of teachers' unions in the United States, forcing a likely AFT/NEA merger to build union solidarity.[37] As state legislatures have become increasingly involved in education policymaking over the past two decades, initiating new policies and programs, teachers' unions have often found themselves in near constant opposition—on the outside looking in. In the case of school choice, despite often-virulent opposition, state legislatures moved rapidly to adopt charter school legislation and voucher proposals are now under consideration in more than half the states. A number of states, including those with powerful teachers' unions, have adopted what the Center for Education Reform calls "strong" charter school laws—laws that allow great flexibility in creating and operating charter schools—including exemptions from collective bargaining agreements and relaxed rules on hiring certified teachers.

It is no accident that the "strongest" charter school laws have the weakest collective bargaining protections. In fact, since charter schools are not exempt from federal health and safety regulations or selected other federal and state provisions, the most common "freedom from red tape" is often the collective bargaining agreement, including exemptions from hiring certified teachers. For example, according to the Center for Education Reform, Rhode Island has one of the weakest, most restrictive charter school laws in the nation, yet the AFT rated it "good."[38] The Rhode Island statute does not allow for private or parochial school conversions—limiting charters to existing public schools. Further, the law requires local approval of two-thirds of the teachers and half the parents; the teachers remain employees of the school district.[39]

Essentially, the "strongest" charter school laws contain the most anti-union provisions. It is difficult, if not impossible, to view the rapid spread of charter schools as in the teachers' unions' best interest or as a victory for organized labor. On the contrary, the rapid spread of what is essentially an anti-union reform suggests that teacher union power is not as strong as it once was. For example, in Michigan, a highly unionized state, Republican Governor John Engler was able to push through a series of anti-union education initiatives. In 1994, Governor Engler "rolled back the scope of [collective] bargaining privileges and limited the ability of teachers to strike. The previous year, he led a successful charge to reform the state's school funding mechanism and to introduce charter schools, a provision loudly opposed by the teachers."[40] Despite vows by the Michigan Education Association to retaliate at the polls, Republicans soon thereafter took control of the legislature and Governor Engler handily won reelection.[41] To cite another example, in the past three mayoral elections in New York City, the candidate backed by the United Federation of Teachers—the largest local affiliate of the American Federation of Teachers—lost each time, including an embarrassing loss in the last mayoral Democratic primary where the union-backed candidate failed even to win his party's own nomination.

Assessing the legislative dynamics of school choice makes one thing clear: the future of school choice lies in coalitional politics—it "will depend on the strength and character of the political coalition that promotes it."[42] These coalitions, particularly those advocating expanded school choice initiatives such as vouchers, are diverse and often fragile—from which small fissures can emerge major breaks—particularly as voucher interests diverge between market-oriented conservatives and equity-oriented minorities.[43] Some fissures have even become apparent within the broad-based coalition supporting charter schools, as the Catholic Church grapples with losing students to urban charter schools.

While vouchers have cleared a significant constitutional hurdle with the U.S. Supreme Court's decision in the *Zelman* case, it would be erroneous to conclude that the battle is over and voucher proponents have won. On the contrary, political battles over vouchers (and other voucher-like proposals such as tuition tax credits) have just begun.[44] The ability of choice proponents (and opponents) to build and sustain effective political coalitions, to manipulate the political process, and to take advantage of quickly opening policy windows will be the deciding factors in legislative battles over choice. These battles will take place in the state legislatures, governors' mansions, courts, the federal government, and in the court of public opinion, as coalitions fight to win the rhetorical war in language games over choice.

Possibly, the school choice movement may mark a new direction in American education toward a more varied, decentralized, and locally controlled system.[45] In this respect, choice is simply *decentralization redux*—and may pose little threat to an educational system that has proven resistant to multiple efforts toward decentralization over the past century. In this respect, charter schools are not a radical reform but merely the logical extension of accountability by way of decentralization. Furthermore, it remains unclear how the drive toward greater parental empowerment and choice will mesh with a more tightly coupled, outcomes-based accountability system that requires ever-greater state and federal oversight.

Finally, despite all the recent attention being paid to school vouchers, the reform remains only one of a number of education issues facing state legislatures, including the significant challenges presented in the federal No Child Left Behind Act of 2001—which may ultimately constitute the greatest federal intrusion (or intervention, depending on one's perspective) into education policy in U.S. history. Faced with a balky economy and increasingly tight fiscal constraints, state lawmakers may find it difficult to maintain current levels of funding for education, much less initiate new programs.[46] Citing sagging tax revenues and growing health care costs, a report released by the National Governors Association and the National Association of State Budget Officers claims that states face their worst fiscal situation since World War II.[47]

Economic constraints may limit the amount of time devoted to voucher initiatives in forthcoming legislative sessions, particularly given the degree of conflict such proposals generate. For example, in Texas, under unified Republican control for the first time since Reconstruction, the legislature is faced with a multibillion-dollar budgetary shortfall. Although the political stars are in alignment for vouchers to come to Texas (as well as in a few other states), the economic climate could not be worse. Texas Governor Rick Perry has made it clear that his top three legislative priorities are: passing a balanced budget with no new taxes, passing medical malpractice legislation, and fixing homeowners' insurance.[48] In a climate of projected (and large) deficits, any choice proposals that require significant expenditures are not likely to find a warm reception in state capitals nationwide. Studies of state education policymaking have found that economic factors play a major role in the nature, scope, and direction of reforms.[49] Economic constraints and an already crowded legislative agenda may keep vouchers from occupying center stage. Furthermore, if Democrats are able to take political advantage of the slumping economy and rising deficits (which to date they have been woefully unable to do), then the next few years may bring about a mild resurgence in Democratic strength that may forestall the push for vouchers.

More likely, the momentum generated by the *Zelman* decision will lead to the adoption of somewhat less controversial reforms such as tuition tax credits or tax deductions, which represent modifications of the existing tax code (and thus draw less public attention), if the budgetary impact of these proposals is limited. While critics deride such schemes as backdoor vouchers, the fact that such plans do not contain the dreaded "V-word" makes them more politically palatable to state policymakers. Another option under consideration by Congress is a proposal to provide federally funded special education vouchers in the reauthorization of the Individuals with Disabilities Education Act (IDEA). Although the outcome of this proposal is very much in doubt, the proposal would not even have been considered had not Republicans taken control of the Senate.

THE BEGINNING OF THE END OR THE END OF THE BEGINNING?

What will be the outcome of the intense legislative battles over school choice—particularly over charter schools and vouchers? Are we witnessing the beginning of the end of public education as a government-run monopoly? Are we at the forefront of a new revolution in education? Or is the revolution already well underway, with the first phase (the evolution from magnet schools to charter schools) completed and the second phase (vouchers) now initiated with the *Zelman* decision? While it may be too early to offer accurate predictions, preliminary indications suggest the outcome of legislative battles over choice will have less impact on public education than either proponents desire or opponents fear. Despite what some analysts call the explosive growth of charter schooling in the United States, after a decade, only a small proportion of students are educated in charter schools. While in some cases the school choice movement has placed much needed pressure on the traditional public school system to reform and better serve the needs of students, in very few instances have charter schools posed much of a threat to the traditional public school system.

Likewise, despite all the hortatory political rhetoric surrounding school vouchers, the public school system, even in cities like Milwaukee and Cleveland, is alive, if not always well. Under public school choice provisions contained in the federal No Child Left Behind Act, students attending schools that fail to demonstrate adequate yearly progress are eligible to attend other nearby public schools, at no cost to the family or student. However, in several states, state and local education officials are actively resisting these new requirements. By revising its criteria, Ohio whittled its list of failing schools from 760 to 212. In Texas and Kentucky, officials waited until after the school year began to release the list of schools to which students could transfer.

In several cities, few eligible students are exercising choice, and some district officials are doing their best to ensure that few students utilize it. In Colorado Springs, less than 2 percent of eligible students have exercised the choice option.[50] In Chicago, where about 125,000 students are eligible to transfer to neighboring public schools, district officials have identified only 2,500 available slots.[51] In Baltimore, about 30,000 students are eligible to transfer; yet district officials claim that just under 200 slots are available at better-performing schools.[52] However, in Fulton County, Georgia, Clark County, Nevada, and Montgomery County, Maryland, less than 3 percent of eligible students are expected to transfer, despite district capacity to accommodate more student transfers.[53] In Pensacola, Florida, state-financed vouchers were made available to 840 families in two failing schools, yet only 58 families opted to use the voucher to send their children to private schools; another 80 families transferred their children to other public schools.[54]

Whether because so few parents are willing to exercise choice, a lack of awareness of choice options, or the limited capacity (and unwillingness) of existing public schools to meet the demand for choice, preliminary evidence suggests that expanded school choice options, even a voucher system, will not lead to a mass exodus from public schools. The sky is not falling, and the old institutional order is not crumbling. Those who think otherwise seriously underestimate the size and resiliency of the public school system (and its supporters). If we are witnessing a revolution, it is a slow-moving, uneven revolution whose outcome is very much in doubt.

The lasting legacy of the push for greater school choice, particularly vouchers, may lay in the ability of reformers to extract greater concessions from the Education Establishment, particularly teachers' unions, which critics have long derided as the most serious obstacle to education reform. For decades, airtight collective bargaining agreements have posed significant barriers to educational reform and change. Now, the very real threat of expanded school choice vis-à-vis vouchers may provide reformers with the leverage necessary to extract significant concessions (such as the end of teacher tenure and beginnings of merit-based pay) from the unions, concessions that would have been unthinkable even a decade earlier. Ultimately, such victories may be the lasting legacy (and real impact) of the school choice movement.

Appendix A

Members of the Pro-Voucher Coalition in Texas

Austin Children's Education Opportunity Foundation
Catholic Church
CEO America
CEO San Antonio
Free Market Committee
National Federation of Independent Businesses
Putting Children First
Texas Justice Foundation
Texas Public Policy Foundation

Appendix B

Members of the Coalition for Public Schools

Advocacy, Inc.
ACLU of Texas
American Association of University Women
American Jewish Committee
American Jewish Congress
Americans United for Separation of Church & State
Association of Texas Professional Educators
Christian Life Commission
Delta Kappa Gamma of Texas
Let Freedom Ring
Parents for Public Schools
Pastors for Peace
People for the American Way Action Fund
Texas AFL-CIO
Texas Association of Community Schools
Texas Association of School Administrators
Texas Association of School Boards
Texas Association of Secondary School Principals
Texas Classroom Teachers Association
Texas Congress of Parents and Teachers
Texas Counseling Association
Texas Elementary Principals and Supervisors Association
Texas Federation of Teachers
Texas Freedom Network
Texas State Teachers Association

Notes

Chapter 1 School Choice and the Politics of Reform

1. Hubert Morken and Jo Renee Formicola note that school choice encompasses a variety of diverse school reform initiatives, including magnet schools and charter schools (public school choice), programs providing public and private scholarships to economically disadvantaged students, public schools managed by for-profit companies, vouchers, tax credits, and home schooling. See Hubert Morken and Jo Renee Formicola, *The Politics of School Choice* (Lanham, MD: Rowman & Littlefield, 1999).
2. Bruce Fuller, Richard F. Elmore, and Gary Orfield, "Policy-Making in the Dark," in Bruce Fuller, Richard F. Elmore, and Gary Orfield, eds., *Who Chooses? Who Loses?* (New York: Teachers College Press, 1996), p. 17.
3. E. Vance Randall, Bruce S. Cooper, and Steven J. Hite, "Understanding the Politics of Research in Education," in Bruce S. Cooper and E. Vance Randall, eds., *Accuracy or Advocacy: The Politics of Research in Education* (Thousand Oaks, CA: Corwin Press, 1999), pp. 1–16.
4. John W. Kingdon, *Agendas, Alternatives, and Public Policy*, 2nd ed. (New York: Longman, 1995).
5. Douglas E. Mitchell, "Educational Politics and Policy: The State Level," in Norman J. Boyan, ed., *Handbook of Research on Educational Administration* (New York: Longman, 1988), p. 453.
6. Pedro Reyes, Lonnie H. Wagstaff, and Lance D. Fusarelli, "Delta Forces: The Changing Fabric of American Society and Education," in Joseph Murphy and Karen Seashore Louis, eds., *Handbook of Research on Educational Administration*, 2nd ed. (San Francisco: Jossey-Bass, 1999), pp. 183–201; Frederick M. Wirt and Michael W. Kirst, *The Political Dynamics of American Education* (Berkeley, CA: McCutchan, 1997).
7. Joel Spring, *Conflict of Interests: The Politics of American Education* (New York: Longman, 1988).
8. Richard P. Nathan, "The Role of the States in American Federalism," in Carl E. Van Horn, ed., *The State of the States*, 3rd ed. (Washington, D.C.: Congressional Quarterly Press, 1996), pp. 13–32.

9. Denis P. Doyle, Bruce S. Cooper, and Roberta Trachtman, eds., *Taking Charge: State Action on School Reform in the 1980s* (Indianapolis, IN: Hudson Institute, 1991).
10. Frederick M. Wirt and Michael W. Kirst, *The Political Dynamics of American Education* (Berkeley, CA: McCutchan, 1997), p. 195.
11. Margaret E. Goertz, "State Education Policy in the 1990s," in C. E. Van Horn, ed., *The State of the States,* 3rd ed. (Washington, D.C.: Congressional Quarterly Press, 1996), pp. 179–208.
12. Jerome T. Murphy, "Progress and Problems: The Paradox of State Reform," in Ann Lieberman and Milbrey W. McLaughlin, eds., *Policy Making in Education* (Chicago: University of Chicago Press, 1982), pp. 195–214.
13. The extent of public support for choice, particularly for vouchers, is a somewhat contentious issue. Pro-voucher organizations such as the Center for Education Reform (CER) have found consistently greater support for choice than have anti-voucher groups such as the United Federation of Teachers and the National Education Association. The CER has been sharply critical of the annual Phi Delta Kappa public opinion poll conducted by the Gallup organization over how the voucher questions are worded in the poll.
14. Joseph P. Viteritti, *Choosing Equality: School Choice, the Constitution, and Civil Society* (Washington, D.C.: The Brookings Institution, 1999).
15. Hubert Morken and Jo Renee Formicola, *The Politics of School Choice* (Lanham, MD: Rowman & Littlefield, 1999), p. 3.
16. Cited in E. Vance Randall, Steven J. Hite, Alan Cheung, and Cheng Biao, *2000 Utah Education Poll—Public Attitudes Towards Public Schools* (Provo, UT: Department of Educational Leadership and Foundations, Brigham Young University, 2001).
17. Linda Jacobson, "Polls Find Growing Support for Publicly Funded Vouchers," *Education Week* 22:1 (September 4, 2002), p. 7.
18. Amy Stuart Wells, *Time to Choose: America at the Crossroads of School Choice Policy* (New York: Hill and Wang, 1993).
19. A. Phillips Brooks, "Texans Support School Vouchers," *Austin American-Statesman* (March 2, 1997), pp. B1, B3.
20. *CER Newswire* 4:32 (August 7, 2002). Available at: http://www.edreform.com.
21. *CER Newswire* 4:30 (July 30, 2002). Available at: http://www.edreform.com.
22. As a guide to policymakers, measures of public support for school vouchers are essentially meaningless, given the general lack of public understanding of the issue, coupled with the simplicity of most public opinion polls. For example, a recent Associated Press poll of 1,011 adults nationwide found that 51 percent favored the idea of vouchers (with 40 percent opposed); however, when asked whether they would still support the idea if it meant taking money from public schools, these same adults opposed

vouchers by a 2-1 margin; see Will Lester, "Voucher Support Fades When Loss to Public Schools is Factored in," *Pittsburgh Post-Gazette* (August 7, 2002), p. A7.

23. As repeated Phi Delta Kappan public opinion polls of U.S. education show, most Americans, particularly those who exercise choice by choosing to live in suburban areas, are quite supportive of public schools.
24. William Ayres, "Navigating a Restless Sea: The Continuing Struggle to Achieve a Decent Education for African-American Youngsters in Chicago," *Journal of Negro Education* 63:1 (1994), pp. 5–18.
25. "Saving Public Education," *The Nation* (February 17, 1997), pp. 17, 18, 20–25.
26. Lance D. Fusarelli, "Leadership in Latino Schools: Challenges for the New Millennium," in Patrick M. Jenlink, ed., *Marching Into a New Millennium: Challenges to Educational Leadership* (Lanham, MD: Scarecrow Press, 2000), pp. 228–238.
27. *CER Newswire* 4:22 (June 11, 2002). Available at: http://www.edreform.com; Pedro Reyes, Lonnie H. Wagstaff, and Lance D. Fusarelli, "Delta Forces: The Changing Fabric of American Society and Education," in Joseph Murphy and Karen Seashore Louis, eds., *Handbook of Research on Educational Administration*, 2nd ed. (San Francisco: Jossey-Bass, 1999), pp. 183–201.
28. Based on average student proficiency in reading from the National Assessment of Educational Progress (NAEP); see National Center for Education Statistics, *Digest of Education Statistics 2001* (Washington, D.C.: U.S. Department of Education, 2002), Table 112, p. 133.
29. National Center for Education Statistics, *Digest of Education Statistics 2001* (Washington, D.C.: U.S. Department of Education, 2002).
30. *The Achiever* 2:1 (January 15, 2003), p. 3.
31. 2000 data.
32. 2000–2001 data.
33. National Center for Education Statistics, *Digest of Education Statistics 2001* (Washington, D.C.: U.S. Department of Education, 2002).
34. Joseph P. Viteritti, "Coming Around on School Choice," *Educational Leadership* (April 2002), pp. 44–47.
35. James G. Cibulka, "Two Eras of Urban Schooling: The Decline of the Old Order and the Emergence of New Organizational Forms," *Education and Urban Society* 29:3 (1997), p. 317.
36. Lance D. Fusarelli, "Reinventing Urban Education in Texas: Charter Schools, Smaller Schools, and the New Institutionalism," *Education and Urban Society* 31:2 (1999), p. 215.
37. Lance D. Fusarelli, "Reinventing Urban Education in Texas: Charter Schools, Smaller Schools, and the New Institutionalism," *Education and Urban Society* 31:2 (1999), p. 215.
38. Joseph Murphy and Catherine Dunn Shiffman, *Understanding and Assessing the Charter School Movement* (New York: Teachers College Press, 2002).

39. Denis P. Doyle, Bruce S. Cooper, and Roberta Trachtman, *Taking Charge: State Action on School Reform in the 1980s* (Indianapolis: Hudson Institute, 1991); House Research Organization, *Charter Schools, Vouchers, and Other School Choice Options*, HRO Publication No. 189 (Austin, June 28, 1994).
40. Erik W. Robelen, "Few Choosing Public School Choice for this Fall," *Education Week* 21:43 (August 7, 2002), pp. 1, 38, 39.
41. Joe Nathan, "Minnesota and the Charter Public School Idea," in Sandra Vergari, ed., *The Charter School Landscape* (Pittsburgh: University of Pittsburgh Press, 2002), pp. 17–31.
42. Carol Smrekar and Ellen Goldring, *School Choice in Urban America: Magnet Schools and the Pursuit of Equity* (New York: Teachers College Press, 1999).
43. Michael Mintrom and David N. Plank, "School Choice in Michigan," in Paul E. Peterson and David E. Campbell, eds., *Charters, Vouchers, and Public Education* (Washington, D.C.: Brookings Institution Press, 2001), pp. 43–58.
44. John Danner and J. C. Bowman, *The Promise and Perils of Charter School Reform*, Tennessee Institute for Public Policy, p. 4. Available at: tnpolicy.org/Education/Eduarch/Charter%20Back.html.
45. Hubert Morken and Jo Renee Formicola, *The Politics of School Choice* (Lanham, MD: Rowman & Littlefield, 1999), p. 4.
46. Catherine Gewertz, "Miami-Dade will Launch Choice Plan," *Education Week* 22:10 (November 6, 2002), pp. 1, 11.
47. Catherine Gewertz, "Miami-Dade will Launch Choice Plan," *Education Week* 22:10 (November 6, 2002), pp. 1, 11.
48. Catherine Gewertz, "Miami-Dade will Launch Choice Plan," *Education Week* 22:10 (November 6, 2002), pp. 1, 11.
49. Catherine Gewertz, "Miami-Dade will Launch Choice Plan," *Education Week* 22:10 (November 6, 2002), p. 11.
50. Michael Mintrom and David N. Plank, "School Choice in Michigan," in Paul E. Peterson and David E. Campbell, eds., *Charters, Vouchers, and Public Education* (Washington, D.C.: Brookings Institution Press, 2001), pp. 43–58.
51. Michael Mintrom and David N. Plank, "School Choice in Michigan," in Paul E. Peterson and David E. Campbell, eds., *Charters, Vouchers, and Public Education* (Washington, D.C.: Brookings Institution Press, 2001), pp. 43–58.
52. Center for Education Reform. Available at: www.edreform.com.
53. CER Press Release (September 17, 2002); Chester E. Finn Jr., Bruno V. Manno, and Gregg Vanourek, "Charter School Accountability: What's a School Board to do?" *American School Board Journal* 187:10 (2000), pp. 42–46.
54. Chester E. Finn, Jr., Bruno V. Manno, and Gregg Vanourek, "Charter Schools: Taking Stock," in Paul E. Peterson and David E. Campbell,

eds., *Charters, Vouchers, and Public Education* (Washington, D.C.: Brookings Institution Press, 2001), pp. 19–42.

55. Joseph Murphy and Catherine Dunn Shiffman, *Understanding and Assessing the Charter School Movement* (New York: Teachers College Press, 2002), p. 34.
56. An excellent and easily accessible, albeit brief, introduction to the Alum Rock experiment may be found in Robert C. Bulman and David L. Kirp, "The Shifting Politics of School Choice," in Stephen D. Sugarman and Frank R. Kemerer, eds., *School Choice and Social Controversy* (Washington, D.C.: Brookings Institution Press, 1999), pp. 40–43.
57. James Brooke, "Denver Minorities Sue for School Vouchers," *Austin American-Statesman* (December 27, 1997), p. A5.
58. Joseph P. Viteritti, "Coming Around on School Choice," *Educational Leadership* (April 2002), p. 46.
59. Milwaukee's voucher program was initiated in 1990; Cleveland's in 1995; Florida's program passed the legislature in 1999.
60. Alan Richard, "Florida Sees Surge in Use of Vouchers," *Education Week* 22:1 (September 4, 2002), pp. 1, 34, 35.
61. Alan Richard, "Florida Sees Surge in Use of Vouchers," *Education Week* 22:1 (September 4, 2002), pp. 1, 34, 35.
62. Alan Richard, "Florida Sees Surge in Use of Vouchers," *Education Week* 22:1 (September 4, 2002), pp. 1, 34, 35.
63. Alan Richard, "Florida Sees Surge in Use of Vouchers," *Education Week* 22:1 (September 4, 2002), pp. 1, 34, 35.
64. Alan Richard, "Florida Sees Surge in Use of Vouchers," *Education Week* 22:1 (September 4, 2002), pp. 1, 34, 35.
65. Hanne B. Mawhinney, "An Interpretive Framework for Understanding the Politics of Policy Change," paper presented at the Annual Meeting of the Canadian Association for Studies in Educational Administration, Calgary, Canada (1994), p. 1.
66. Peter A. Hall, "Policy Paradigms, Social Learning, and the State: The Case of Economic Policymaking in Britain," *Comparative Politics* 25:3 (1993), pp. 275–296; Tim L. Mazzoni, "The Changing Politics of State Education Policymaking: A 20-year Minnesota Perspective," *Educational Evaluation and Policy Analysis* 15:4 (1993), pp. 357–379; Gerald E. Sroufe, "Politics of Education at the Federal Level," in Jay D. Scribner and Donald H. Layton, eds., *The Study of Educational Politics* (London: Falmer Press, 1995), pp. 75–88.
67. Hanne B. Mawhinney, "An Interpretive Framework for Understanding the Politics of Policy Change," paper presented at the Annual Meeting of the Canadian Association for Studies in Educational Administration, Calgary, Canada (1994), pp. 1–2.
68. Gerald E. Sroufe, "Politics of Education at the Federal Level," in Jay D. Scribner and Donald H. Layton, eds., *The Study of Educational Politics* (London: Falmer Press, 1995), p. 79.

69. Tim L. Mazzoni, "The Changing Politics of State Education Policymaking: A 20-year Minnesota Perspective," *Educational Evaluation and Policy Analysis* 15:4 (1993), pp. 357–379.
70. Gaye Tuchman, "Historical Social Science," in Norman K. Denzin and Yvonna S. Lincoln, eds., *Handbook of Qualitative Research* (Thousand Oaks, CA: Sage, 1994), p. 306.
71. Richard E. Neustadt and Ernest R. May, *Thinking in Time: The Uses of History for Decision-makers* (New York: The Free Press, 1986), p. xxi.
72. Richard E. Neustadt and Ernest R. May, *Thinking in Time: The Uses of History for Decision-makers* (New York: The Free Press, 1986), p. xxi.
73. Ann Majchrzak, *Methods for Policy Research* (Newbury Park, CA: Sage, 1984), p. 20.
74. James J. Scheurich, "Policy Archaeology: A New Policy Studies Methodology," *Journal of Education Policy* 9:4 (1994), p. 313.
75. Donald R. Warren, "A Past for the Present," in Donald R. Warren, ed., *History, Education, and Public Policy* (Berkeley, CA: McCutchan, 1978), p. 13.
76. Bryan C. Hassel, *The Charter School Challenge* (Washington, D.C.: Brookings Institution Press, 1999), p. 17.
77. Daniel J. Elazar, *American Federalism: A View From the States* (New York: Crowell, 1972); Daniel J. Elazar, *The American Mosaic: The Impact of Space, Time, and Culture on American Politics* (Boulder, CO: Westview Press, 1994); Richard J. Ellis, *American Political Cultures* (New York: Oxford University Press, 1993); Dennis J. Coyle and Richard J. Ellis, eds., *Politics, Policy, and Culture* (Boulder, CO: Westview Press, 1994).
78. James G. March and Johan P. Olsen, *Rediscovering Institutions: The Organization Basis of Politics* (New York: The Free Press, 1989); Theda Skocpol, *Protecting Soldiers and Mothers: The Political Origins of Social Policy in the United States* (Cambridge: Harvard University Press, 1992); Stephen Skowronek, *Building a New American State: The Expansion of National Administrative Capacities, 1877–1920* (Cambridge: Cambridge University Press, 1982); Sven Steinmo, Kathleen Thelen, and Frank Longstreth, eds., *Structuring Politics: Historical Institutionalism in Comparative Perspective* (Cambridge: Cambridge University Press, 1992).
79. David B. Truman, *The Governmental Process: Political Interests and Public Opinion* (New York: Knopf, 1964); Hank C. Jenkins-Smith and Paul A. Sabatier, "Evaluating the Advocacy Coalition Framework," *Journal of Public Policy* 14:2 (1994), pp. 175–203; Paul A. Sabatier, "Policy Change Over a Decade or More," in Paul A. Sabatier and Hank C. Jenkins-Smith, eds., *Policy Change and Learning: An Advocacy Coalition Approach* (Boulder, CO: Westview Press, 1993), pp. 13–39.
80. Frans L. Leeuw, Ray C. Rist, and Richard C. Sonnichsen, eds., *Can Governments Learn? Comparative Perspectives on Evaluation and Organizational Learning* (New Brunswick, NJ: Transaction Publishers, 1994); Peter A. Hall, "Policy Paradigms, Social Learning, and the

State: The Case of Economic Policymaking in Britain," *Comparative Politics* 25:3 (1993), pp. 275–296; Ray C. Rist, "The Preconditions for Learning: Lessons From the Public Sector," in Frans L. Leeuw, Ray C. Rist, and Richard C. Sonnichsen, eds., *Can Governments Learn? Comparative Perspectives on Evaluation and Organizational Learning* (New Brunswick, NJ: Transaction Publishers, 1994), pp. 189–205.

81. Carol H. Weiss, ed., *Using Social Research in Public Policymaking* (Lexington, MA: D. C. Heath and Company, 1977); Carol H. Weiss, "Ideology, Interests, and Information: The Basis of Policy Decisions," in Daniel Callahan and Bruce Jennings, eds., *Ethics, the Social Sciences and Policy Analysis* (New York: Plenum Press, 1983), pp. 213–245.
82. Peter Senge, *The Fifth Discipline: The Art and Practice of the Learning Organization* (New York: Doubleday, 1990).

Chapter 2 Cultural Dynamics: Political Culture and Language in Policymaking

1. E. E. Schattschneider, *The Semi-Sovereign People: A Realist's View of Democracy in America* (Hinsdale, IL: Dryden Press, 1960), p. 68.
2. Daniel J. Elazar, *American Federalism*, 3rd ed. (New York: Harper and Row, 1984); Daniel J. Elazar, *The American Mosaic: The Impact of Space, Time, and Culture on American Politics* (Boulder, CO: Westview Press, 1994), p. 3.
3. D. J. Elkins and R. E. B. Simeon, "A Cause in Search of its Effect, or What Does Political Culture Explain?" *Comparative Politics* 11:2 (1979), p. 128.
4. D. J. Elkins and R. E. B. Simeon, "A Cause in Search of its Effect, or What Does Political Culture Explain?" *Comparative Politics* 11:2 (1979), p. 128.
5. David A. Rochefort and Roger W. Cobb, eds., *The Politics of Problem Definition* (Lawrence: University Press of Kansas, 1994).
6. Christopher J. Bosso, "The Contextual Bases of Problem Definition," in David A. Rochefort and Roger W. Cobb, eds., *The Politics of Problem Definition* (Lawrence: University Press of Kansas, 1994), p. 199.
7. D. J. Elkins and R. E. B. Simeon, "A Cause in Search of its Effect, or What Does Political Culture Explain?" *Comparative Politics* 11:2 (1979), p. 131.
8. Daniel J. Elazar, *The American Mosaic: The Impact of Space, Time, and Culture on American Politics* (Boulder, CO: Westview Press, 1994), p. 3.
9. Daniel J. Elazar, *The American Mosaic: The Impact of Space, Time, and Culture on American Politics* (Boulder, CO: Westview Press, 1994), p. 277.
10. Daniel J. Elazar, *The American Mosaic: The Impact of Space, Time, and Culture on American Politics* (Boulder, CO: Westview Press, 1994), p. 231.
11. Richard J. Ellis, *American Political Cultures* (New York: Oxford University Press, 1993), p. 151.

12. Daniel J. Elazar, *The American Mosaic: The Impact of Space, Time, and Culture on American Politics* (Boulder, CO: Westview Press, 1994), p. 234.
13. Thomas James, "State Authority and the Politics of Educational Change," *Review of Research in Education* 17 (1991), p. 195.
14. Tim L. Mazzoni, "The Changing Politics of State Education Policymaking: A 20-year Minnesota Perspective," *Educational Evaluation and Policy Analysis* 15:4 (1993), pp. 357–379.
15. Tim L. Mazzoni, "The Changing Politics of State Education Policymaking: A 20-year Minnesota Perspective," *Educational Evaluation and Policy Analysis* 15:4 (1993), pp. 357–379.
16. Catherine Marshall, Douglas Mitchell, and Frederick Wirt, *Culture and Education Policy in the American States* (New York: Falmer Press, 1989); Douglas Mitchell, Catherine Marshall, and Frederick Wirt, "Building a Taxonomy of State Education Policies," *Peabody Journal of Education* 62 (1985), pp. 7–47; Frederick Wirt, Douglas Mitchell, and Catherine Marshall, "Culture and Education Policy: Analyzing Values in State Policy Systems," *Educational Evaluation and Policy Analysis* 10:4 (1988), pp. 271–284.
17. Donal M. Sacken and Marcello Medina, Jr., "Investigating the Context of State-Level Policy Formation: A Case Study of Arizona's Bilingual Education Legislation," *Educational Evaluation and Policy Analysis* 12:4 (1990), pp. 389–402.
18. Maenette K. P. Benham and Ronald H. Heck, "Political Culture and Policy in a State-Controlled Educational System: The Case of Educational Politics in Hawaii," *Educational Administration Quarterly* 30:4 (1994), pp. 419–450.
19. Peter H. Garland and S. V. Martorana, "The Interplay of Political Culture and Participant Behavior in Political Action to Enact Significant State Community College Legislation," *Community College Review* 16:2 (1988), pp. 30–43.
20. C. Elaine Freeman, *Missouri and Oklahoma: A Comparative Study of State Higher Education Policy and Political Culture*, paper presented at the Annual Meeting of the American Educational Research Association, San Francisco, CA (April 20–24, 1992).
21. Lorn S. Foster, "Political Culture: A Determinant of Public Regardingness among School Board Members," *Urban Education* 18:1 (1983), pp. 29–39.
22. Belinda L. Pustka, *Relationship of Political Culture Orientation to Policy Decisions Made at the School District Level*, unpublished doctoral dissertation, The University of Texas at Austin (1996).
23. Jay D. Scribner and Lance D. Fusarelli, "Rethinking the Nexus Between Religion and Political Culture: Implications for Educational Policy," *Education and Urban Society* 28:3 (1996), pp. 289–290; Albert Marten, *Operationalizing Political Culture: An Application to Texas Independent School Districts*, unpublished doctoral dissertation, The University of

Texas at Austin (1993); Jay D. Scribner and Albert Marten, *Measuring the Political Culture of School Districts: The Texas Case*, paper presented at the annual meeting of the American Educational Research Association, New Orleans, LA (April 1994); See also the May 1996 special issue of *Education and Urban Society* 28:3 devoted to the role of religion in the politics of education.

24. Joe Nathan, "Minnesota and the Charter Public School Idea," in Sandra Vergari, ed., *The Charter School Landscape* (Pittsburgh: University of Pittsburgh Press, 2002), pp. 17–31.
25. This analysis, of course, depends on viewing charter schools as a movement to enhance equity in education. Opponents of charter schools, on the other hand, argue that charter schools exacerbate inequities in the public school system. A number of scholars who have studied charter school effects have found inequities in several areas, including services for special education and limited English proficient (LEP) students; See Sandra Vergari, ed., *The Charter School Landscape* (Pittsburgh: University of Pittsburgh Press, 2002).
26. Joseph P. Viteritti, *Choosing Equality: School Choice, the Constitution, and Civil Society* (Washington, D.C.: Brookings Institution Press, 1999), pp. 62–63.
27. Joseph P. Viteritti, *Choosing Equality: School Choice, the Constitution, and Civil Society* (Washington, D.C.: Brookings Institution Press, 1999), p. 63.
28. Daniel J. Elazar, *The American Mosaic: The Impact of Space, Time, and Culture on American Politics* (Boulder, CO: Westview Press, 1994), p. 236.
29. Ellis Katz, "Pennsylvania," in Susan Fuhrman and Alan Rosenthal, eds., *Shaping Education Policy in the States* (Washington, D.C.: Institute for Educational Leadership, 1981), p. 17.
30. Ellis Katz, "Pennsylvania," in Susan Fuhrman and Alan Rosenthal, eds., *Shaping Education Policy in the States* (Washington, D.C.: Institute for Educational Leadership, 1981), p. 31.
31. Ellis Katz, "Pennsylvania," in Susan Fuhrman and Alan Rosenthal, eds., *Shaping Education Policy in the States* (Washington, D.C.: Institute for Educational Leadership, 1981), p. 34.
32. Frederick M. Hess and Robert Maranto, "Letting a Thousand Flowers (and Weeds) Bloom: The Charter Story in Arizona," in Sandra Vergari, ed., *The Charter School Landscape* (Pittsburgh: University of Pittsburgh Press, 2002), p. 54.
33. Frederick M. Hess and Robert Maranto, "Letting a Thousand Flowers (and Weeds) Bloom: The Charter Story in Arizona," in Sandra Vergari, ed., *The Charter School Landscape* (Pittsburgh: University of Pittsburgh Press, 2002), p. 55.
34. Daniel J. Elazar, *The American Mosaic: The Impact of Space, Time, and Culture on American Politics* (Boulder, CO: Westview Press, 1994).
35. Linda L. M. Bennett, *Symbolic State Politics: Education Funding in Ohio, 1970–1980* (New York: Peter Lang, 1983).

36. Linda L. M. Bennett, *Symbolic State Politics: Education Funding in Ohio, 1970–1980* (New York: Peter Lang, 1983), p. 138.
37. Bryan C. Hassel, *The Charter School Challenge* (Washington, D.C.: Brookings Institution Press, 1999), p. 28.
38. Bryan C. Hassel, *The Charter School Challenge* (Washington, D.C.: Brookings Institution Press, 1999), p. 24.
39. Robert N. Bellah, Richard Madsen, William M. Sullivan, Ann Swidler, and Steven M. Tipton, *Habits of the Heart: Individualism and Commitment in American Life* (New York: Harper & Row, 1985); Robert N. Bellah, Richard Madsen, William M. Sullivan, Ann Swidler, and Steven M. Tipton, *The Good Society* (New York: Vintage Books, 1991).
40. John W. Kingdon, *Agendas, Alternatives, and Public Policies*, 2nd ed. (Boston: Addison-Wesley, 1995).
41. Ludwig Wittgenstein, *Philosophical Grammar*, R. Rhees, ed., A. Kenny, trans. (Berkeley: University of California Press, 1974); Murray Edelman, *The Symbolic Uses of Politics* (Urbana: University of Illinois Press, 1964); Murray Edelman, *Constructing the Political Spectacle* (Chicago: University of Chicago Press, 1988).
42. At first glance, such differences may appear to be readily apparent. However, by carefully listening to policymakers, educators, or even doctoral students discuss such terms, it becomes clear that a significant amount of definitional ambiguity exists in these terms. The politics of defining the terminology we use plays a pivotal role in how we view the merits of each approach.
43. Murray Edelman, *The Symbolic Uses of Politics* (Urbana: University of Illinois Press, 1964).
44. Murray Edelman, *Constructing the Political Spectacle* (Chicago: University of Chicago Press, 1988), p. 103.
45. Richard Rorty, *Contingency, Irony, and Solidarity* (Cambridge, MA: Cambridge University Press, 1989).
46. Richard S. Prawat and Penelope L. Peterson, "Social Constructivist Views of Learning," in J. Murphy and K. Seashore-Louis, eds., *Handbook of Research on Educational Administration*, 2nd ed. (San Francisco: Jossey Bass, 1999), p. 211.
47. Murray Edelman, *The Symbolic Uses of Politics* (Urbana: University of Illinois Press, 1964), pp. 119–120.
48. Deborah A. Stone, *Policy Paradox and Political Reason* (Glenview, IL: Scott, Foresman, 1988), p. 200.
49. Murray Edelman, *Constructing the Political Spectacle* (Chicago: University of Chicago Press, 1988).
50. M. P. Petracca, "Issue Definitions, Agenda-Building, and Policymaking," *Policy Currents* 2 (1992), p. 1.
51. William H. Riker, *The Art of Political Manipulation* (New Haven: Yale University Press, 1986).

52. E. E. Schattschneider, *The Semi-Sovereign People: A Realist's View of Democracy in America* (Hinsdale, IL: Dryden Press, 1960), p. 4.
53. E. E. Schattschneider, *The Semi-Sovereign People: A Realist's View of Democracy in America* (Hinsdale, IL: Dryden Press, 1960), p. 69.
54. Murray Edelman, *Constructing the Political Spectacle* (Chicago: University of Chicago Press, 1988), pp. 103–104.
55. Daniel J. Elazar, *The American Mosaic: The Impact of Space, Time, and Culture on American Politics* (Boulder, CO: Westview Press, 1994), p. 5; B. S. Fennimore, *Talk Matters: Refocusing the Language of Public Schooling* (New York: Teachers College Press, 2000).
56. Daniel J. Elazar, *The American Mosaic: The Impact of Space, Time, and Culture on American Politics* (Boulder, CO: Westview Press, 1994), p. 5.
57. Hilary Putnam, *Words and Life* (Cambridge, MA: Harvard University Press, 1994).
58. Noreen B. Garman and Patricia C. Holland, "The Rhetoric of School Reform Reports: Sacred, Skeptical, and Cynical Interpretations," in Rick Ginsberg and David N. Plank, eds., *Commissions, Reports, Reforms, and Educational Policy* (Westport, CT: Praeger, 1995), pp. 101–117.
59. Peter Westen, *Speaking of Equality* (Princeton, NJ: Princeton University Press, 1990).
60. Peter Westen, *Speaking of Equality* (Princeton, NJ: Princeton University Press, 1990), p. 261.
61. E. E. Schattschneider, *The Semi-Sovereign People: A Realist's View of Democracy in America* (Hinsdale, IL: Dryden Press, 1960), p. 69.
62. Richard S. Prawat and Penelope L. Peterson, "Social Constructivist Views of Learning," in J. Murphy and K. Seashore-Louis, eds., *Handbook of Research on Educational Administration*, 2nd ed. (San Francisco: Jossey Bass, 1999), p. 214.
63. G. Brand, *The Essential Wittgenstein*, R. E. Innis, trans. (New York: Basic Books, 1979); Ludwig Wittgenstein, *Philosophical Grammar*, R. Rhees, ed., A. Kenny, trans. (Berkeley: University of California Press, 1974).
64. William H. Riker, *The Strategy of Rhetoric* (New Haven: Yale University Press, 1996), p. 4.
65. Peter Westen, *Speaking of Equality* (Princeton, NJ: Princeton University Press, 1990), p. 257.
66. Noreen B. Garman and Patricia C. Holland, "The Rhetoric of School Reform Reports: Sacred, Skeptical, and Cynical Interpretations," in Rick Ginsberg and David N. Plank, eds., *Commissions, Reports, Reforms, and Educational Policy* (Westport, CT: Praeger, 1995), p. 106.
67. Noreen B. Garman and Patricia C. Holland, "The Rhetoric of School Reform Reports: Sacred, Skeptical, and Cynical Interpretations," in Rick Ginsberg and David N. Plank, eds., *Commissions, Reports, Reforms, and Educational Policy* (Westport, CT: Praeger, 1995), p. 106.
68. Ann Norton, *Republic of Signs: Liberal Theory and American Popular Culture* (Chicago: University of Chicago Press, 1993), p. 1.

69. David B. Tyack, *The One Best System: A History of American Urban Education* (Cambridge: Harvard University Press, 1974); Elizabeth Hansot and David B. Tyack, "A Usable Past: Using History in Educational Policy," in Ann Lieberman and Milbrey W. McLaughlin, eds., *Policy Making in Education* (Chicago: University of Chicago Press, 1982), pp. 5–10.
70. Ira Katznelson and Margaret Weir, *Schooling for All: Class, Race, and the Decline of the Democratic Ideal* (New York: Basic Books, 1985).
71. Ira Katznelson and Margaret Weir, *Schooling for All: Class, Race, and the Decline of the Democratic Ideal* (New York: Basic Books, 1985).
72. Joe Christie, "Charter Schools Provide Autonomy, Accountability," *Austin American-Statesman* (April 8, 1994), p. A11.
73. Kent Grusendorf, "School Choice Deserves a Chance," *Austin American-Statesman* (March 30, 1994), p. A13.
74. Kent Grusendorf, "School Choice Deserves a Chance," *Austin American-Statesman* (March 30, 1994), p. A13.
75. K. Shannon, "Bush Endorses Flexible Approach for Public Schools," *Austin American-Statesman* (July 2, 1996), p. B3.
76. During the campaign, both Richards and Bush called for a return to local control of education and a reversal of the trend toward state centralization; A. Phillips Brooks, "Texas Doing Homework on Charter Schools Concept," *Austin American-Statesman* (February 20, 1994), p. A1; A. Phillips Brooks, "Education Reform Unites Opponents," *Austin American-Statesman* (May 8, 1994), pp. B1, B7.
77. "Steady As She Goes," *Education Week* 16 (January 22, 1997), pp. 214–216.
78. House Research Organization, *Charter Schools, Vouchers, and Other School Choice Options* (HRO Publication No. 189) (Austin, 1994).
79. A. Phillips Brooks, "Texas Doing Homework on Charter Schools Concept," *Austin American-Statesman* (February 20, 1994), p. A17.
80. A. Phillips Brooks, "Texas Doing Homework on Charter Schools Concept," *Austin American-Statesman* (February 20, 1994), p. A17.
81. A. Phillips Brooks, "Education to Remain a Priority, Bush Says," *Austin American-Statesman* (November 13, 1996), p. B3.
82. "Vouchers Are No Cure," *Austin American-Statesman* (March 27, 1997), p. A14.
83. A. Phillips Brooks, "Limitations Abound in Education Vouchers," *Austin American-Statesman* (April 30, 1995), p. A15.
84. A. Phillips Brooks, "Limitations Abound in Education Vouchers," *Austin American-Statesman* (April 30, 1995), p. A15.
85. "Vouchers Are No Cure," *Austin American-Statesman* (March 27, 1997), p. A14.
86. A. Phillips Brooks, "School Voucher Proposal Omitted by House Panel," *Austin American-Statesman* (March 31, 1995), p. B3.
87. A. Phillips Brooks, "Voucher Plans Open New Battle for Schools," *Austin American-Statesman* (March 13, 1997), p. B1.

88. A. Phillips Brooks, "Voucher Opponents Say They Can Block Bill in Senate," *Austin American-Statesman* (March 26, 1997), p. B1.
89. A. Phillips Brooks, "Education Reform Unites Opponents," *Austin American-Statesman* (May 8, 1994), p. B7.
90. S.B. 1 authorized the State Board of Education (SBOE) to grant open-enrollment charters to schools operated by a public, private, or independent institution of higher education, a non-profit organization, or a governmental entity. These schools are considered part of the public school system and are subject to the same regulations as home-rule charter schools.
91. J. Berls, "Parents Seek to Create Minority Charter Schools," *Austin American-Statesman* (March 21, 1996), pp. B1, B5.
92. A. Phillips Brooks, "Texans Support School Vouchers," *Austin American-Statesman* (March 2, 1997), p. B1.
93. House Research Organization, *Charter Schools, Vouchers, and Other School Choice Options* (HRO Publication No. 189) (Austin, 1994), p. 1.
94. A. Phillips Brooks, "House Votes Down Vouchers, Adopts Values Measures," *Austin American-Statesman* (May 7, 1995), p. B2.
95. Putting Children First is a coalition of two hundred business executives and civic groups led by Mansour and former state senator Joe Christie.
96. A. Phillips Brooks, "Voucher Plans Open New Battle for Schools," *Austin American-Statesman* (March 13, 1997), p. B1.
97. Kent Grusendorf, "School Choice Deserves a Chance," *Austin American-Statesman* (March 30, 1994), p. A13.
98. A. Phillips Brooks, "Conferees Settle Issues on Education," *Austin American-Statesman* (May 17, 1995), p. B6.
99. A. Phillips Brooks, "Conferees Settle Issues on Education," *Austin American-Statesman* (May 17, 1995), pp. B1, B6.
100. A. Phillips Brooks, "Area School Districts Rejecting Vouchers," *Austin American-Statesman* (August 12, 1995), p. B6.
101. A. Phillips Brooks, "Choice Programs Won't Fix Public Schools, Study Says," *Austin American-Statesman* (March 30, 1994), pp. B1–B2.
102. A. Phillips Brooks, "Choice Programs Won't Fix Public Schools, Study Says," *Austin American-Statesman* (March 30, 1994), p. B2.
103. "Education Hearing Covers Charters, Segregation Fears," *Austin American-Statesman* (March 2, 1995), p. B3.
104. A. Phillips Brooks, "Limitations Abound in Education Vouchers," *Austin American-Statesman* (April 30, 1995), p. A15.
105. A. Phillips Brooks, "Charters, Other Plans Worry Minority Lawmakers," *Austin American-Statesman* (May 12, 1995), p. B3.
106. A. Phillips Brooks, "Charters, Other Plans Worry Minority Lawmakers," *Austin American-Statesman* (May 12, 1995), p. B3.
107. A. Phillips Brooks, "Minorities Criticize Compromise Bill," *Austin American-Statesman* (May 25, 1995), p. B3.
108. A. Phillips Brooks, "Vouchers Opposed by Many Minority Lawmakers," *Austin American-Statesman* (June 6, 1996), p. B6.

109. A. Phillips Brooks, "Hispanic Groups to Grade Lawmakers," *Austin American-Statesman* (April 16, 1997), p. B3.
110. A. Phillips Brooks, "Panel Supports School Vouchers," *Austin American-Statesman* (April 17, 1997), p. B8.
111. Texas is a relatively poor state, with wide discrepancies between the rich and poor (which the political culture does little to discourage). Texas is also a state of minorities, with the fourth largest minority population in the United States. Accordingly, among the policy elite, elected representatives must be sensitive to the impact of policies on people of color. Fulfilling their role as trustees, policy elites tend to shy away from proposals that threaten to further widen the gap between rich and poor, especially when key choice advocates come from the Republican Right.
112. Murray Edelman, *Constructing the Political Spectacle* (Chicago: University of Chicago Press, 1988), pp. 103–104.
113. Deborah A. Stone, *Policy Paradox and Political Reason* (Glenview, IL: Scott, Foresman, 1988); William H. Riker, *The Art of Political Manipulation* (New Haven: Yale University Press, 1986); William H. Riker, *The Strategy of Rhetoric* (New Haven: Yale University Press, 1996).
114. Christopher J. Bosso, "The Contextual Bases of Problem Definition," in David A. Rochefort and Roger W. Cobb, eds., *The Politics of Problem Definition* (Lawrence: University Press of Kansas, 1994), pp. 182–203.
115. Ann Bastian, "Is Public School 'Choice' a Viable Alternative?" in D. Levine, R. Lowe, B. Peterson, and R. Tenorio, eds., *Rethinking Schools: An Agenda for Change* (New York: The Free Press, 1995), p. 205.
116. Stefan D. Haag, Rex C. Peebles, and Gary A. Keith, *Texas Politics and Government: Ideas, Institutions, and Policies* (New York: Longman, 1997).
117. Linda M. McNeil, "The Politics of Texas School Reform," in William L. Boyd and Charles T. Kerchner, eds., *The Politics of Excellence and Choice in Education* (New York: Falmer Press, 1988), p. 203.
118. Frederick M. Hess and Robert Maranto, *Letting a Thousand Flowers (and Weeds) Bloom: The Charter Story in Arizona*, paper presented at the Annual Meeting of the American Educational Research Association. New Orleans, LA, 2000.
119. Martin Rein and Donald Schon, "Reframing Policy Discourse," in Frank Fischer and John Forester, eds., *The Argumentative Turn in Policy Analysis and Planning* (Durham, NC: Duke University Press, 1993), p. 156.
120. Noreen B. Garman and Patricia C. Holland, "The Rhetoric of School Reform Reports: Sacred, Skeptical, and Cynical Interpretations," in Rick Ginsberg and David N. Plank, eds., *Commissions, Reports, Reforms, and Educational Policy* (Westport, CT: Praeger, 1995), pp. 101–117.
121. Peter J. Shelly, "Opponents Tough on Ridge School Plan," *Pittsburgh Post-Gazette* (May 28, 1999), p. D1.

122. Peter J. Shelly, "Opponents Tough on Ridge School Plan," *Pittsburgh Post-Gazette* (May 28, 1999), p. D1.
123. Patti Lather, *Getting Smart: Feminist Research and Pedagogy With/in the Postmodern* (New York: Routledge, 1991); Patti Lather, "Troubling Clarity: The Politics of Accessible Language," *Harvard Educational Review* 66:3 (1996), pp. 525–545.
124. Patti Lather, *Getting Smart: Feminist Research and Pedagogy With/in the Postmodern* (New York: Routledge, 1991), pp. 11–12.
125. Patti Lather, "Troubling Clarity: The Politics of Accessible Language," *Harvard Educational Review* 66:3 (1996), p. 528.
126. Cleo H. Cherryholmes, *Power and Criticism: Poststructural Investigations in Education* (New York: Teachers College Press, 1988), p. 35.
127. Lilia I. Bartolome and Donaldo P. Macedo, "Dancing with Bigotry: The Poisoning of Racial and Ethnic Identities," *Harvard Educational Review* 67:2 (1997), p. 233.
128. Christopher J. Bosso, "The Contextual Bases of Problem Definition," in David A. Rochefort and Roger W. Cobb, eds., *The Politics of Problem Definition* (Lawrence: University Press of Kansas, 1994), pp. 182–203.
129. Hubert Morken and Jo R. Formicola, *The Politics of School Choice* (Lanham, MD: Rowman & Littlefield, 1999); C. R. Sauter, *Charter Schools: A New Breed of Public Schools* (Oak Brook, IL: The North Central Regional Educational Laboratory, 1993).
130. Paul C. Bauman, *Governing Education: Public Sector Reform or Privatization* (Boston: Allyn and Bacon, 1996), p. 122.
131. Robert C. Bulman and David L. Kirp, "The Shifting Politics of School Choice," in Stephen D. Sugarman and Frank R. Kemerer, eds., *School Choice and Social Controversy* (Washington, D.C.: Brookings Institution Press, 1999), p. 60.
132. Bella Rosenberg of the American Federation of Teachers argues that some charter schools clearly represent an effort to establish private schools within a public system. I agree and argue that the distinction made between the public and private spheres is itself a type of language game—one which has heretofore been dominated by the success of policymakers in maintaining the public/private dichotomy, particularly the dominant view within the Educational Establishment of the public sphere as "good" and the private sphere as "bad." I posit that the distinction between charter schools and voucher plans is less clear than the rhetoric of policy debates would suggest. Taebel and Brenner (1994) note, for example, that "school choice has been used in many forms [in Texas] ranging from alternative education programs, magnet programs, inter- and intradistrict transfers, tax credits and vouchers which can be applied toward tuition and other educational services." See D. A. Taebel and C. T. Brenner, eds., *Privatizing Public Education: The Texas Experience* (Arlington: Institute of Urban Studies & Center for Economic Development Research and Service, 1994), p. 99. It is clear from

observations of the daily practice of schooling that the two spheres are mixed to a significantly greater extent than is commonly believed or publicly acknowledged.

133. Sheldon S. Wolin, *The Presence of the Past: Essays on the State and the Constitution* (Baltimore: Johns Hopkins University Press, 1989).
134. Michelle Fine, "The 'Public' in Public Schools: The Social Construction/Constriction of Moral Communities," *Journal of Social Issues* 46:1 (1990), p. 114.
135. William H. Riker, *The Art of Political Manipulation* (New Haven: Yale University Press, 1986); E. E. Schattschneider, *The Semi-Sovereign People: A Realist's View of Democracy in America* (Hinsdale, IL: Dryden Press, 1960); Deborah A. Stone, *Policy Paradox and Political Reason* (Glenview, IL: Scott, Foresman, 1988); Deborah A. Stone, *Policy Paradox: The Art of Political Decisionmaking* (New York: W. W. Norton, 1997).
136. Charles D. Elder and Roger W. Cobb, *The Political Uses of Symbols* (New York: Longman, 1983).
137. David A. Rochefort and Roger W. Cobb, "Problem Definition: An Emerging Perspective," in David A. Rochefort and Roger W. Cobb, eds., *The Politics of Problem Definition* (Lawrence: University Press of Kansas, 1994), p. 27.
138. Richard S. Prawat and Penelope L. Peterson, "Social Constructivist Views of Learning," in J. Murphy and K. Seashore-Louis, eds., *Handbook of Research on Education Administration*, 2nd ed. (San Francisco: Jossey Bass, 1999), pp. 203–226.
139. David Bloor, *Wittgenstein: A Social Theory of Knowledge* (New York: Columbia University Press, 1983), as cited in Prawat and Peterson, p. 214.
140. Richard Rorty, *Contingency, Irony, and Solidarity* (Cambridge, MA: Cambridge University Press, 1989).

CHAPTER 3 INSTITUTIONAL DYNAMICS: THE POWER OF STRUCTURE

1. B. Guy Peters, *Institutional Theory in Political Science* (London: Pinter, 1999).
2. For an excellent application of neo-institutional theory to education, see Robert L. Crowson, William L. Boyd, and Hanne B. Mawhinney, eds., *The Politics of Education and the New Institutionalism: Reinventing the American School* (London: Falmer Press, 1996).
3. James G. March and Johan P. Olsen, *Rediscovering Institutions: The Organization Basis of Politics* (New York: The Free Press, 1989).
4. James G. March and Johan P. Olsen, *Rediscovering Institutions: The Organization Basis of Politics* (New York: The Free Press, 1989), p. 1.
5. R. Kent Weaver and Bert A. Rockman, "Assessing the Effects of Institutions," in R. Kent Weaver and Bert A. Rockman, eds., *Do Institutions Matter? Government Capabilities in the United States and Abroad* (Washington, D.C.: The Brookings Institution, 1993), pp. 1–41.

6. Kathleen Thelen and Sven Steinmo, "Historical Institutionalism in Comparative Politics," in Sven Steinmo, Kathleen Thelen, and Frank Longstreth, eds., *Structuring Politics: Historical Institutionalism in Comparative Analysis* (Cambridge: Cambridge University Press, 1992), p. 1.
7. Kathleen Thelen and Sven Steinmo, "Historical Institutionalism in Comparative Politics," in Sven Steinmo, Kathleen Thelen, and Frank Longstreth, eds., *Structuring Politics: Historical Institutionalism in Comparative Analysis* (Cambridge: Cambridge University Press, 1992), p. 5.
8. James G. Cibulka, "Two Eras of Schooling: The Decline of the Old Order and the Emergence of New Organizational Forms," *Education and Urban Society* 29:3 (1997), p. 318.
9. Peter Senge, *The Fifth Discipline: The Art and Practice of the Learning Organization* (New York: Doubleday, 1990), p. 43.
10. John W. Kingdon, *Agendas, Alternatives, and Public Policies* (Boston: Little, Brown and Company, 1984), p. 217.
11. David B. Robertson and Dennis R. Judd, *The Development of American Public Policy: The Structure of Policy Restraint* (Glenview, IL: Scott, Foresman and Company, 1989).
12. Theda Skocpol, *Protecting Soldiers and Mothers: The Political Origins of Social Policy in the United States* (Cambridge: Harvard University Press, 1992), p. 41.
13. G. John Ikenberry, "Conclusion: An Institutional Approach to American Foreign Economic Policy," in G. John Ikenberry, David A. Lake, and Michael Mastanduno, eds., *The State and American Foreign Economic Policy* (Ithaca, NY: Cornell University Press, 1988), pp. 219–243.
14. James G. March and Johan P. Olsen, *Rediscovering Institutions: The Organization Basis of Politics* (New York: The Free Press, 1989), p. 4.
15. Stephen D. Krasner, "Approaches to the State: Alternative Conceptions and Historical Dynamics," *Comparative Politics* 16:2 (1984), p. 228.
16. James G. March and Johan P. Olsen, "The New Institutionalism: Organizational Factors of Political Life," *American Political Science Review* 78 (1984), p. 739.
17. Peter J. Katzenstein, "Introduction: Domestic and International Forces and Strategies of Foreign Economic Policy," in Peter J. Katzenstein, ed., *Between Power and Plenty: Foreign Economic Policies of Advanced Industrial States* (Madison: University of Wisconsin Press, 1978), pp. 3–22.
18. James G. March and Johan P. Olsen, *Rediscovering Institutions: The Organization Basis of Politics* (New York: The Free Press, 1989), p. 17.
19. Kathleen Thelen and Sven Steinmo, "Historical Institutionalism in Comparative Politics," in Sven Steinmo, Kathleen Thelen, and Frank Longstreth, eds., *Structuring Politics: Historical Institutionalism in Comparative Analysis* (Cambridge: Cambridge University Press, 1992), pp. 1–32.

20. Kathleen Thelen and Sven Steinmo, "Historical Institutionalism in Comparative Politics," in Sven Steinmo, Kathleen Thelen, and Frank Longstreth, eds., *Structuring Politics: Historical Institutionalism in Comparative Analysis* (Cambridge: Cambridge University Press, 1992), p. 9.
21. James G. March and Johan P. Olsen, *Rediscovering Institutions: The Organization Basis of Politics* (New York: The Free Press, 1989); L. A. Pal, *Public Policy Analysis: An Introduction*, 2nd ed. (Scarborough, ON: Nelson Canada, 1992).
22. Stephen D. Krasner, "Approaches to the State: Alternative Conceptions and Historical Dynamics," *Comparative Politics* 16:2 (1984), p. 228.
23. James G. Cibulka, "Two Eras of Schooling: The Decline of the Old Order and the Emergence of New Organizational Forms," *Education and Urban Society* 29:3 (1997), pp. 318–319.
24. Kathleen Thelen and Sven Steinmo, "Historical Institutionalism in Comparative Politics," in Sven Steinmo, Kathleen Thelen, and Frank Longstreth, eds., *Structuring Politics: Historical Institutionalism in Comparative Analysis* (Cambridge: Cambridge University Press, 1992), p. 8.
25. James G. March and Johan P. Olsen, *Rediscovering Institutions: The Organization Basis of Politics* (New York: The Free Press, 1989).
26. Nancy C. Roberts and Paula J. King, *Transforming Public Policy: Dynamics of Policy Entrepreneurship and Innovation* (San Francisco: Jossey-Bass, 1996), p. 17.
27. Thomas James, "State Authority and the Politics of Educational Change," *Review of Research in Education* 17 (1991), p. 177.
28. Theda Skocpol, *Protecting Soldiers and Mothers: The Political Origins of Social Policy in the United States* (Cambridge: Harvard University Press, 1992), p. 42.
29. James G. March and Johan P. Olsen, *Rediscovering Institutions: The Organization Basis of Politics* (New York: The Free Press, 1989), p. 1.
30. E. E. Schattschneider, *The Semi-Sovereign People: A Realist's View of Democracy in America* (Hinsdale, IL: Dryden Press, 1960), p. 70.
31. E. E. Schattschneider, *The Semi-Sovereign People: A Realist's View of Democracy in America* (Hinsdale, IL: Dryden Press, 1960), p. 69.
32. Theda Skocpol, *Protecting Soldiers and Mothers: The Political Origins of Social Policy in the United States* (Cambridge: Harvard University Press, 1992), p. 54.
33. Douglas M. Abrams, *Conflict, Competition, or Cooperation? Dilemmas of State Education Policymaking* (Albany: State University of New York Press. 1993), p. 29.
34. Hanne B. Mawhinney, "An Interpretive Framework for Understanding the Politics of Policy Change," paper presented at the Annual Meeting of the Canadian Association for Studies in Educational Administration, Calgary, Canada (1994), p. 19.

35. Peter Evans, Dietrich Rueschmeyer, and Theda Skocpol, eds., *Bringing the State Back In* (New York: Cambridge University Press, 1985).
36. Peter A. Hall, "Policy Paradigms, Social Learning, and the State: The Case of Economic Policymaking in Britain," *Comparative Politics* 25:3 (1993), p. 275.
37. Stephen Skowronek, *Building a New American State: The Expansion of National Administrative Capacities, 1877–1920* (Cambridge: Cambridge University Press, 1982), p. 20.
38. Thomas James, "State Authority and the Politics of Educational Change," *Review of Research in Education* 17 (1991), p. 173.
39. Thomas James, "State Authority and the Politics of Educational Change," *Review of Research in Education* 17 (1991), p. 190.
40. Thomas James, "State Authority and the Politics of Educational Change," *Review of Research in Education* 17 (1991), p. 187; Tim L. Mazzoni, "The Changing Politics of State Education Policymaking: A 20-year Minnesota Perspective," *Educational Evaluation and Policy Analysis* 15:4 (1993), pp. 357–379; Alan Rosenthal, "The Emerging Legislative Role in Education," *Compact* 11:1 (1977), pp. 2–4.
41. Carl E. Van Horn, ed., *The State of the States*, 3rd ed. (Washington, D.C.: Congressional Quarterly Inc., 1996).
42. Peter J. Katzenstein, "Introduction: Domestic and International Forces and Strategies of Foreign Economic Policy," in Peter J. Katzenstein, ed., *Between Power and Plenty: Foreign Economic Policies of Advanced Industrial States* (Madison: University of Wisconsin Press, 1978), p. 19.
43. Peter J. Katzenstein, "Conclusion: Domestic Structures and Strategies of Foreign Economic Policy," in Peter J. Katzenstein, ed., *Between Power and Plenty: Foreign Economic Policies of Advanced Industrial States* (Madison: University of Wisconsin Press, 1978), p. 306.
44. Theda Skocpol, *Protecting Soldiers and Mothers: The Political Origins of Social Policy in the United States* (Cambridge: Harvard University Press, 1992), p. 58.
45. David B. Robertson, "The Return to History and the New Institutionalism in American Political Science," *Social Science History* 17:1 (1993), p. 19.
46. Theda Skocpol, *Protecting Soldiers and Mothers: The Political Origins of Social Policy in the United States* (Cambridge: Harvard University Press, 1992).
47. Theda Skocpol, *Protecting Soldiers and Mothers: The Political Origins of Social Policy in the United States* (Cambridge: Harvard University Press, 1992), p. 58.
48. Theda Skocpol, *Protecting Soldiers and Mothers: The Political Origins of Social Policy in the United States* (Cambridge: Harvard University Press, 1992), p. 58.
49. Richard E. Neustadt and Ernest R. May, *Thinking in Time: The Uses of History for Decision-makers* (New York: The Free Press, 1986), p. xxi.

50. Elaine K. Swift and David W. Brady, "Out of the Past: Theoretical and Methodological Contributions of Congressional History," *PS: Political Science & Politics* 24:1 (1991), p. 61.
51. Margaret Weir, "Ideas and the Politics of Bounded Innovation," in Sven Steinmo, Kathleen Thelen, and Frank Longstreth, eds., *Structuring Politics: Historical Institutionalism in Comparative Analysis* (Cambridge: Cambridge University Press, 1992), pp. 188–216.
52. Richard E. Neustadt and Ernest R. May, *Thinking in Time: The Uses of History for Decision-makers* (New York: The Free Press, 1986).
53. Kathleen Thelen and Sven Steinmo, "Historical Institutionalism in Comparative Politics," in Sven Steinmo, Kathleen Thelen, and Frank Longstreth, eds., *Structuring Politics: Historical Institutionalism in Comparative Analysis* (Cambridge: Cambridge University Press, 1992), pp. 11–12.
54. Kathleen Thelen and Sven Steinmo, "Historical Institutionalism in Comparative Politics," in Sven Steinmo, Kathleen Thelen, and Frank Longstreth, eds., *Structuring Politics: Historical Institutionalism in Comparative Analysis* (Cambridge: Cambridge University Press, 1992), p. 13.
55. David B. Robertson, "The Return to History and the New Institutionalism in American Political Science," *Social Science History* 17:1 (1993), p. 24.
56. Mavis M. Reeves, "The States as Polities: Reformed, Reinvigorated, Resourceful," *Annals of the American Academy of Political and Social Science*, 509 (1990), p. 88; Alan Rosenthal, "The Emerging Legislative Role in Education," *Compact* 11:1 (1977), pp. 2–4; Alan Rosenthal, "The New Legislature: Better or Worse and for Whom?" in Thad L. Beyle, ed., *State Government: CQ's Guide to Current Issues and Activities, 1987–88* (Washington, D.C.: Congressional Quarterly Inc., 1989), pp. 69–70.
57. David B. Robertson and Dennis R. Judd, *The Development of American Public Policy: The Structure of Policy Restraint* (Glenview, IL: Scott, Foresman and Company, 1989), p. 10.
58. Denis P. Doyle, Bruce S. Cooper, and Roberta Trachtman, *Taking Charge: State Action on School Reform in the 1980s* (Indianapolis, IN: Hudson Institute, 1991), p. 1.
59. Joseph Murphy, "The Educational Reform Movement of the 1980s: A Comprehensive Analysis," in Joseph Murphy, ed., *The Educational Reform Movement of the 1980s: Perspectives and Cases* (Berkeley, CA: McCutchan, 1990), p. 21.
60. William A. Firestone, "Continuity and Incrementalism After All: State Responses to the Excellence Movement," in Joseph Murphy, ed., *The Educational Reform Movement of the 1980s: Perspectives and Cases* (Berkeley, CA: McCutchan, 1990), pp. 146; Michael W. Kirst and Stephen A. Somers, "California Educational Interest Groups: Collective Action as a Logical Response to Proposition 13," *Education and Urban Society* 13:2 (1981), pp. 235–256.

61. Tim L. Mazzoni, "The Changing Politics of State Education Policymaking: A 20-year Minnesota Perspective," *Educational Evaluation and Policy Analysis* 15:4 (1993), pp. 357–379; Sue Wells Weaver and Terry G. Geske, "Educational Policy-making in the State Legislature: Legislator as Policy Expert," paper presented at the Annual Meeting of the American Educational Research Association, New York, New York, 1996.
62. Sue Wells Weaver and Terry G. Geske, "Educational Policy-making in the State Legislature: Legislator as Policy Expert," paper presented at the Annual Meeting of the American Educational Research Association, New York, New York, 1996.
63. Stephen Skowronek, *Building a New American State: The Expansion of National Administrative Capacities, 1877–1920* (Cambridge: Cambridge University Press, 1982), p. 12.
64. Tim L. Mazzoni, "The Changing Politics of State Education Policymaking: A 20-year Minnesota Perspective," *Educational Evaluation and Policy Analysis* 15:4 (1993), p. 365.
65. Christopher J. Bosso, "The Contextual Bases of Problem Definition," in David A. Rochefort and Roger W. Cobb, eds., *The Politics of Problem Definition* (Lawrence: University Press of Kansas, 1994), pp. 182–203; Tim L. Mazzoni, "The Changing Politics of State Education Policymaking: A 20-year Minnesota Perspective," *Educational Evaluation and Policy Analysis* 15:4 (1993), pp. 357–379; David A. Rochefort and Roger W. Cobb, eds., *The Politics of Problem Definition* (Lawrence: University Press of Kansas, 1994).
66. Hubert Morken and Jo Renee Formicola, *The Politics of School Choice* (Lanham, MD: Rowman & Littlefield, 1999).
67. Hubert Morken and Jo Renee Formicola, *The Politics of School Choice* (Lanham, MD: Rowman & Littlefield, 1999), p. 21.
68. Hubert Morken and Jo Renee Formicola, *The Politics of School Choice* (Lanham, MD: Rowman & Littlefield, 1999), p. 74.
69. Michael Mintrom and David N. Plank, "School Choice in Michigan," in Paul E. Peterson and David E. Campbell, eds., *Charters, Vouchers, and Public Education* (Washington, D.C.: Brookings Institution Press, 2001), p. 47.
70. *Zelman v. Simmons-Harris*, 436 U.S. 2002.
71. The Center for Education Reform notes that the federal government has a long history of creating special programs for the poor that permit public funds to be used in private and religious institutions—such as government vouchers for daycare at private and parochial facilities, federally funded Pell grants and National Direct Student Loans in higher education, and G.I. benefits—all of which can be used to attend private and religious institutions (*CER Newswire* 2:44, December 12, 2000); see Lawrence D. Weinberg, Bruce S. Cooper, and Lance D. Fusarelli, "Education Vouchers for Religious Schools: Legal and Social Justice Perspectives," *Religion & Education* 27:1 (2000), pp. 34–42 for a more detailed discussion of this issue.

72. Center for Education Reform, *CER Newswire—Special Edition* 4:25 (June 28, 2002). Available at: http://www.edreform.com.
73. John Gehring, "Voucher Battles Head to State Capitals," *Education Week* 21:42 (July 10, 2002), pp. 1, 24, 25.
74. *CER Newswire* 3:15 (April 10, 2001). Available at: http://www.edreform.com.
75. A. Phillips Brooks, "Senate Approves Education Overhaul," *Austin American-Statesman* (March 28, 1995), p. A6.
76. Frank Reeves, "Vouching for Vouchers," *Pittsburgh Post-Gazette* (March 31, 1999), p. B1.
77. The initiative allows citizens to propose a law by petition and submit it to voters for approval. Twenty-four states and the District of Columbia have the initiative process. A referendum gives voters a chance to approve or disapprove legislation proposed by the state legislature; House Research Organization, *Charter Schools, Vouchers, and Other School Choice Options*, HRO Publication No. 189 (Austin, June 28, 1994).
78. J. J. Miller, "Why School Choice Lost," *Wall Street Journal* (November 4, 1993), p. A8.
79. *CAPE Update* (November 8, 2000). Available at: http://www.capenet.org.
80. Michael E. Jewell and Samuel C. Patterson, *The Legislative Process in the United States* (New York: Random House, 1966), p. 138.
81. Virginia Baxt and Liane Brouillette, "The State, the Lobbyists, and Special Education Policies in Schools: A Case Study of Decision Making in Texas," *Journal of School Leadership* 9, p. 136.
82. A. Phillips Brooks, "House Approval Sends Education Bill to Negotiating Table," *Austin American-Statesman* (May 8, 1995), p. B3.
83. In their analysis of special education policies in S.B. 1, Baxt and Brouillette point out that the omnibus bill allowed special education inclusion policies to be inserted (almost stealthily) and passed with little notice (see Virginia Baxt and Liane Brouillette, "The State, the Lobbyists, and Special Education Policies in Schools: A Case Study of Decision Making in Texas," *Journal of School Leadership* 9, pp. 125–159).
84. A. Phillips Brooks, "Overhaul of Code Proposed," *Austin American-Statesman* (February 15, 1995), p. B3.
85. Virginia Baxt and Liane Brouillette, "The State, the Lobbyists, and Special Education Policies in Schools: A Case Study of Decision Making in Texas," *Journal of School Leadership* 9, p. 148.
86. Texas Classroom Teachers Association, "75th session," (1998), Available: www.tcta.org/legover.htm.
87. David Bible, "Texas Legislature Rejects Vouchers, 'Parental Rights,'" *The North Texas Activist!* Source: www.flash.net/~lbartley/au/activist/act0103/sessio75.htm.
88. Clay Robison, "V-Word Sidestepped in Austin and D.C.," *Houston Chronicle* (January 27, 2001), p. A1.

89. L. Harmon Zeigler and H. van Dalen, "Interest Groups in the States," in Herbert Jacob and Kenneth N. Vines, eds., *Politics in the American States* (Boston: Little, Brown and Company, 1971), pp. 122–160.
90. L. Harmon Zeigler and H. van Dalen, "Interest Groups in the States," in Herbert Jacob and Kenneth N. Vines, eds., *Politics in the American States* (Boston: Little, Brown and Company, 1971), pp. 122–160.
91. L. Harmon Zeigler and M. A. Baer, *Lobbying: Interaction and Influence in American State Legislatures* (Belmont, CA: Wadsworth, 1969).
92. Citizens Conference on State Legislatures, *The Sometime Governments: A Critical Study of the 50 American Legislatures* (Kansas City, MO, 1973), p. 103.
93. John Gehring, "Voucher Battles Head to State Capitals," *Education Week* 21:42 (July 10, 2002), pp. 1, 24, 25.
94. John Gehring, "Voucher Battles Head to State Capitals," *Education Week* 21:42 (July 10, 2002), p. 24.
95. Lance D. Fusarelli, "The Political Economy of Gubernatorial Elections: Implications for Education Policy," *Educational Policy* 16:1, pp. 139–160.
96. Lance D. Fusarelli, "The Political Economy of Gubernatorial Elections: Implications for Education Policy," *Educational Policy* 16:1, p. 156.
97. Erik W. Robelen, "Ruling Gives Second Wind to Capitol Hill Voucher Advocates," *Education Week* 21:42 (July 10, 2002), p. 27.
98. This example also highlights the power of committee chairs in shaping education legislation; Chris Sheridan, "Horse-Trading Ahead on Schools," *The Plain Dealer* (November 10, 1996), p. H2.
99. Joseph Murphy and Catherine Dunn Shiffman, *Understanding and Assessing the Charter School Movement* (New York: Teachers College Press, 2002), p. 46.
100. Stefan D. Haag, Gary A. Keith, and Rex C. Peebles, *Texas Politics and Government: Ideas, Institutions, and Policies,* 3rd ed. (New York: Longman, 2003); J. A. Schlesinger, "The Politics of the Executive," in Herbert Jacob and Kenneth N. Vines, eds., *Politics in the American States* (Boston: Little, Brown and Company, 1971), pp. 210–237.
101. The framers were reacting in part to the abuses of Governor E. J. Davis during Reconstruction. Davis is considered by many historians to be one of the most corrupt governors in U.S. history.
102. By comparison, in Pennsylvania, the governor's cabinet meets weekly.
103. Stuart Eskenazi, "Democrat Help Vital for Bush," *Austin American-Statesman* (November 13, 1994), p. A8.
104. Stuart Eskenazi, "Democrat Help Vital for Bush," *Austin American-Statesman* (November 13, 1994), p. A8.
105. Lance D. Fusarelli, "The Political Economy of Gubernatorial Elections: Implications for Education Policy," *Educational Policy* 16:1 (2002), pp. 139–160.
106. Stuart Eskenazi, "Democrat Help Vital for Bush," *Austin American-Statesman* (November 13, 1994), p. A8.

107. Stuart Eskenazi, "Democrat Help Vital for Bush," *Austin American-Statesman* (November 13, 1994), p. A8.
108. Stuart Eskenazi, "Democrat Help Vital for Bush," *Austin American-Statesman* (November 13, 1994), p. A8.
109. Stuart Eskenazi, "Democrat Help Vital for Bush," *Austin American-Statesman* (November 13, 1994), p. A8.
110. Linda L. M. Bennett, *Symbolic State Politics: Education Funding in Ohio, 1970–1980* (New York: Peter Lang, 1983), p. 130.
111. Linda L. M. Bennett, *Symbolic State Politics: Education Funding in Ohio, 1970–1980* (New York: Peter Lang, 1983).
112. Sandy Theis and Mark Skertic, "Budget Boosts Schools, Senate GOP Offers More for Pupils, Buildings," *Cincinnati Enquirer* (May 21, 1997), p. A1.
113. Michael Hawthorne, "GOP's Finan Leads Ohio Senate with his Knuckles Bare," *Cincinnati Enquirer* (May 29, 1997), p. B1.
114. Michael Hawthorne, "GOP's Finan Leads Ohio Senate with his Knuckles Bare," *Cincinnati Enquirer* (May 29, 1997), p. B1.
115. To this day, the lieutenant governors of Georgia, Missouri, and Indiana wield similar power; Citizens Conference on State Legislatures, *The Sometime Governments: A Critical Study of the 50 American Legislatures* (Kansas City, MO, 1973).
116. B. Wear, "Texas Legislature '95: Changes Are Few But Could Be Profound," *Austin American-Statesman* (January 8, 1995), p. D5.
117. In fact, 1997 was his last session as lieutenant governor, after which he retired from public service.
118. Ellis Katz, "Pennsylvania," in Susan Fuhrman and Alan Rosenthal, eds., *Shaping Education Policy in the States* (Washington, D.C.: Institute for Educational Leadership, 1981), p. 34.
119. Ellis Katz, "Pennsylvania," in Susan Fuhrman and Alan Rosenthal, eds., *Shaping Education Policy in the States* (Washington, D.C.: Institute for Educational Leadership, 1981), pp. 20–21.
120. The Calendars Committee has been known to bury bills members of the committee (or the Speaker) do not support.
121. "Fighting Vouchers in Texas: The 1999 Legislative Battle." Available at: http://www.tfn.org/issues/vouchers/1999legfight.htm.
122. "Fighting Vouchers in Texas: The 1999 Legislative Battle." Available at: http://www.tfn.org/issues/vouchers/1999legfight.htm.
123. Due to the Two-Thirds Rule.
124. A. Phillips Brooks, "Panel Supports School Vouchers," *Austin American-Statesman* (April 17, 1997), pp. B1, B8.
125. A. Phillips Brooks, "Voucher Opponents Say They Can Block Bill in Senate," *Austin American-Statesman* (March 26, 1997), p. B1.
126. R. Kent Weaver and Bert A. Rockman, "Assessing the Effects of Institutions," in R. Kent Weaver and Bert A. Rockman, eds., *Do Institutions Matter? Government Capabilities in the United States and Abroad* (Washington, D.C.: The Brookings Institution, 1993), p. 33.

127. George Strong, "Will Vouchers be the Killer Bees of 1999?" (April 17, 1999) available at: http://www.political.com/analysis-arc/0244.html.
128. "Senators Vow to Block Any Voucher Bill That Would Rob Texas' Public Schools of Funds," source: www.groups.yahoo.com/group/FEMNET/message/92.
129. George Strong, "Will Vouchers Be the Killer Bees of 1999?" (April 17, 1999) available at: http://www.political.com/analysis-arc/0244.html.
130. A. Phillips Brooks, "Panel Supports School Vouchers," *Austin American-Statesman* (April 17, 1997), pp. B1, B8.
131. David Bible, "Texas Legislature Rejects Vouchers, 'Parental Rights,'" *The North Texas Activist!* Source: www.flash.net/~lbartley/au/activist/act0103/sessio75.htm.
132. Clay Robison, "7 Democratic Senators Vow to Fight Voucher Proposal," *Houston Chronicle* (March 26, 1997), p. A25.
133. Ken Herman, "Legislature Moving on Job One," *Austin American-Statesman* (April 20, 1997), p. B2.
134. A. Phillips Brooks, "Panel Supports School Vouchers," *Austin American-Statesman* (April 17, 1997), p. B1.
135. Karen Nicholson, "Legislative Update," source: www.midland.k12.tx.us/pta/legisad.htm.
136. In many respects, by cutting down substantially on the number of bills considered by the Senate each session, the Two-Thirds Rule performs a function not unlike the role of plea bargaining in the criminal justice system. Were it not for the plea bargaining system, where most cases are settled out of court and without juries, the entire criminal justice system would quickly grind to a halt, with long periods between arrest and trial. Similarly, the Two-Thirds Rule cuts down substantially on the workload of the Senate, with far fewer bills coming up for debate on the Senate floor. Given an extremely short legislative session, such procedural rules are invaluable, if not overly democratic.
137. In deference to the power of the lieutenant governor, in nearly every interview Bullock was referred to by the title "Governor Bullock"; only once during any of the interviews was he referred to by his actual title of lieutenant governor. This use of language reflects not only Bullock's formal institutional power but his personal power as well. Lieutenant governor Bullock was an institution unto himself in Texas politics—a legend in state politics—feared, respected, and admired by both friends and enemies.
138. Erik W. Robelen, "Ruling Gives Second Wind to Capitol Hill Voucher Advocates," *Education Week* 21:42 (July 10, 2002), p. 27.
139. Erik W. Robelen, "Ruling Gives Second Wind to Capitol Hill Voucher Advocates," *Education Week* 21:42 (July 10, 2002), p. 27.
140. Mary Beth Lane, "'School Choice' Foes Told to Write to Fight Plan," *The Plain Dealer* (March 1, 1995), p. B5.

141. Thomas Suddes, "School Vouchers Face House Debate Thursday; 3 Democratic Attacks on Plan for Cleveland Shot Down," *The Plain Dealer* (April 4, 1995), p. B4.
142. Virginia Baxt and Liane Brouillette, "The State, the Lobbyists, and Special Education Policies in Schools: A Case Study of Decision Making in Texas," *Journal of School Leadership* 9, pp. 125–159.
143. "Fighting Vouchers in Texas: The 1999 Legislative Battle." Available at: http://www.tfn.org/issues/vouchers/1999legfight.htm.
144. "Fighting Vouchers in Texas: The 1999 Legislative Battle." Available at: http://www.tfn.org/issues/vouchers/1999legfight.htm.
145. "Fighting Vouchers in Texas: The 1999 Legislative Battle." Available at: http://www.tfn.org/issues/vouchers/1999legfight.htm.
146. Michael Mintrom, *Policy Entrepreneurs and School Choice* (Washington, D.C.: Georgetown University Press, 2000), p. 65.
147. Hubert Morken and Jo Renee Formicola, *The Politics of School Choice* (Lanham, MD: Rowman & Littlefield, 1999), p. 86.
148. Robert C. Bulman and David L. Kirp, "The Shifting Politics of School Choice," in Stephen D. Sugarman and Frank R. Kemerer, eds., *School Choice and Social Controversy* (Washington, D.C.: Brookings Institution Press, 1999), p. 50.
149. Robert C. Bulman and David L. Kirp, "The Shifting Politics of School Choice," in Stephen D. Sugarman and Frank R. Kemerer, eds., *School Choice and Social Controversy* (Washington, D.C.: Brookings Institution Press, 1999), pp. 56–57.
150. Robert C. Bulman and David L. Kirp, "The Shifting Politics of School Choice," in Stephen D. Sugarman and Frank R. Kemerer, eds., *School Choice and Social Controversy* (Washington, D.C.: Brookings Institution Press, 1999), pp. 56–57.
151. Robert C. Bulman and David L. Kirp, "The Shifting Politics of School Choice," in Stephen D. Sugarman and Frank R. Kemerer, eds., *School Choice and Social Controversy* (Washington, D.C.: Brookings Institution Press, 1999), p. 57.
152. Frederick M. Hess and Robert Maranto, "Letting a Thousand Flowers (and Weeds) Bloom: The Charter Story in Arizona," in Sandra Vergari, ed., *The Charter School Landscape* (Pittsburgh: University of Pittsburgh Press, 2002), p. 56.
153. Reconstruction ended in 1876.
154. Stefan D. Haag, Gary A. Keith, and Rex C. Peebles, *Texas Politics and Government: Ideas, Institutions, and Policies*, 3rd ed. (New York: Longman, 2003).
155. Republican Bill Clements was the first, elected in 1978.
156. A. Phillips Brooks, "Bush Election Revives Controversial Initiatives," *Austin American-Statesman* (November 11, 1994), p. A8.
157. A. Phillips Brooks, "Bush Election Revives Controversial Initiatives," *Austin American-Statesman* (November 11, 1994), p. A8.

158. A. Phillips Brooks, "Bush Election Revives Controversial Initiatives," *Austin American-Statesman* (November 11, 1994), p. A8.
159. A. Phillips Brooks, "Bush Election Revives Controversial Initiatives," *Austin American-Statesman* (November 11, 1994), p. A8.
160. Ellis Katz, "Pennsylvania," in Susan Fuhrman and Alan Rosenthal, eds., *Shaping Education Policy in the States* (Washington, D.C.: Institute for Educational Leadership, 1981), p. 17.
161. Ellis Katz, "Pennsylvania," in Susan Fuhrman and Alan Rosenthal, eds., *Shaping Education Policy in the States* (Washington, D.C.: Institute for Educational Leadership, 1981), p. 17.
162. Frank Reeves and Peter J. Shelly, "Day Care Fight Imperils Voucher Plan," *Pittsburgh Post-Gazette* (April 14, 1999), p. B1.
163. Frank Reeves and Peter J. Shelly, "Day Care Fight Imperils Voucher Plan," *Pittsburgh Post-Gazette* (April 14, 1999), p. B1.
164. Pataki's role as policy entrepreneur in pushing New York's charter school bill through a recalcitrant state legislature is discussed more fully in chapter 4.
165. *Statistical Abstract of the United States, 1999* (Washington, D.C.: U.S. Bureau of the Census, 1999).
166. Bryan C. Hassel, *The Charter School Challenge* (Washington, D.C.: Brookings Institution Press, 1999).
167. Bruce S. Cooper and E. Vance Randall, *Critics Stop Vouchers in Their Tracks*, paper presented at the Education Between State and Civil Society Conference, Boston, MA, 1997, p. 45.
168. Douglas M. Abrams, *Conflict, Competition, or Cooperation? Dilemmas of State Education Policymaking* (Albany: State University of New York Press, 1993); Kathleen Thelen and Sven Steinmo, "Historical Institutionalism in Comparative Politics," in Sven Steinmo, Kathleen Thelen, and Frank Longstreth, eds., *Structuring Politics: Historical Institutionalism in Comparative Analysis* (Cambridge: Cambridge University Press, 1992), pp. 1–32; David B. Truman, *The Governmental Process: Political Interests and Public Opinion* (New York: Knopf, 1964).
169. Hanne B. Mawhinney, "An Interpretive Framework for Understanding the Politics of Policy Change," paper presented at the Annual Meeting of the Canadian Association for Studies in Educational Administration, Calgary, Canada (1994), p. 19.

CHAPTER 4 INTEREST GROUP DYNAMICS: THE POWER OF COALITIONS

1. Paul A. Sabatier, "Toward Better Theories of the Policy Process," *PS: Political Science & Politics* 24:2 (1991), p. 148.
2. Hank C. Jenkins-Smith and Paul A. Sabatier, "Evaluating the Advocacy Coalition Framework," *Journal of Public Policy* 14:2 (1994), p. 179.

3. Daniel McCool, ed., *Public Policy Theories, Models, and Concepts* (Englewood Cliffs, NJ: Prentice Hall, 1995), pp. 251–389.
4. Paul A. Sabatier and Hank C. Jenkins-Smith, eds., *Policy Change and Learning: An Advocacy Coalition Approach* (Boulder, CO: Westview Press, 1993).
5. Paul A. Sabatier and Hank C. Jenkins-Smith, eds., *Policy Change and Learning: An Advocacy Coalition Approach* (Boulder, CO: Westview Press, 1993), p. 5.
6. Paul A. Sabatier, "Policy Change Over a Decade or More," in Paul A. Sabatier and Hank C. Jenkins-Smith, eds., *Policy Change and Learning: An Advocacy Coalition Approach* (Boulder, CO: Westview Press, 1993), p. 19.
7. Joseph Stewart, Jr., "Policy Models and Equal Educational Opportunity," *PS: Political Science & Politics* 24:2 (1991), p. 171.
8. Traditional systems theory incorporates such processes into the analytical framework (see Easton, 1965). As such, advocacy coalition models share elements of this approach. In fact, it should be obvious that much of the intellectual groundwork of advocacy coalition models rests upon traditional systems theory. However, systems theory is not really a testable or verifiable theory of policy change so much as a laundry list of factors affecting policy development and change.
9. Kevin W. Hula, *Lobbying Together: Interest Group Coalitions in Legislative Politics* (Washington, D.C.: Georgetown University Press, 1999), p. 2.
10. Paul A. Sabatier, "An Advocacy Coalition Framework of Policy Change and the Role of Policy-oriented Learning Therein," *Policy Sciences* 21 (1988), pp. 129–168.
11. Frank R. Baumgartner and Bryan D. Jones, "Agenda Dynamics and Policy Subsystems," *Journal of Politics* 53:4 (1991), pp. 1044–1074.
12. Martin Burlingame and Terry G. Geske, "State Politics and Education: An Examination of Selected Multiple-state Case Studies," *Educational Administration Quarterly* 15:2 (1979), p. 71.
13. Hanne B. Mawhinney, "An Interpretive Framework for Understanding the Politics of Policy Change," Ph.D. Diss., University of Ottawa, 1993, p. 412.
14. Tim L. Mazzoni, "The Changing Politics of State Education Policymaking: A 20-year Minnesota Perspective," *Educational Evaluation and Policy Analysis* 15:4 (1993), pp. 357–379.
15. Tim L. Mazzoni, "The Changing Politics of State Education Policymaking: A 20-year Minnesota Perspective," *Educational Evaluation and Policy Analysis* 15:4 (1993), p. 375.
16. Tim L. Mazzoni, "The Changing Politics of State Education Policymaking: A 20-year Minnesota Perspective," *Educational Evaluation and Policy Analysis* 15:4 (1993), p. 377.

17. Tim L. Mazzoni, "The Changing Politics of State Education Policymaking: A 20-year Minnesota Perspective," *Educational Evaluation and Policy Analysis* 15:4 (1993), p. 377.
18. Robert E. Feir, "Political and Social Roots of Education Reform: A Look at the States in the Mid-1980s," paper presented at the Annual Meeting of the American Educational Research Association, San Francisco, CA, 1995.
19. Robert E. Feir, "Political and Social Roots of Education Reform: A Look at the States in the Mid-1980s," paper presented at the Annual Meeting of the American Educational Research Association, San Francisco, CA, 1995, p. 29.
20. Jane H. Karper and William Lowe Boyd, "Interest Groups and the Changing Environment of State Educational Policymaking: Developments in Pennsylvania," *Educational Administration Quarterly* 24:1, p. 28.
21. John W. Kingdon, *Agendas, Alternatives, and Public Policy*, 2nd ed. (New York: Longman, 1995).
22. Michael Mintrom, "Policy Entrepreneurs and the Diffusion of Innovation," *American Journal of Political Science* 41:3 (1997), p. 738; Michael Mintrom, *Policy Entrepreneurs and School Choice* (Washington, D.C.: Georgetown University Press, 2000). See also John W. Kingdon, *Agendas, Alternatives, and Public Policy*, 2nd ed. (New York: Longman, 1995).
23. Hubert Morken and Jo Renee Formicola, *The Politics of School Choice* (Lanham, MD: Rowman & Littlefield, 1999), p. 42.
24. Hubert Morken and Jo Renee Formicola, *The Politics of School Choice* (Lanham, MD: Rowman & Littlefield, 1999), p. 43.
25. Pedro Reyes, Lonnie H. Wagstaff, and Lance D. Fusarelli, "Delta Forces: The Changing Fabric of American Society and Education," in Joseph Murphy and Karen Seashore Louis, eds., *Handbook of Research on Educational Administration*, 2nd ed. (San Francisco: Jossey-Bass, 1999), pp. 183–201.
26. House Research Organization, *Charter Schools, Vouchers, and Other School Choice Options*, HRO Publication No. 189 (Austin, June 28, 1994), p. 1.
27. A. Phillips Brooks, "Texans Support School Vouchers," *Austin American-Statesman* (March 2, 1997), p. B1.
28. A. Phillips Brooks, "Choice Programs Won't Fix Public Schools, Study Says," *Austin American-Statesman* (March 30, 1994), p. B2.
29. A. Phillips Brooks, "Vouchers Opposed by Many Minority Lawmakers," *Austin American-Statesman* (June 6, 1996), p. B6.
30. A. Phillips Brooks, "Charters, Other Plans Worry Minority Lawmakers," *Austin American-Statesman* (May 12, 1995), p. B3.
31. Kent Grusendorf, "School Choice Deserves a Chance," *Austin American-Statesman* (March 30, 1994), p. A13.

32. A. Phillips Brooks, "Conferees Settle Issues on Education," *Austin American-Statesman* (May 17, 1995), p. B6.
33. Kathy Walt, "Group Pushes School Voucher Program," *Houston Chronicle* (March 4, 1997), p. A17; Mansour is also chair of CEO America and a member of the Texas Public Policy Foundation.
34. R. G. Ratcliffe, "Conservative Liberal With His Offerings," *Houston Chronicle* (September 21, 1997), pp. A1, A21.
35. Kathy Walt, "Group Pushes School Voucher Program," *Houston Chronicle* (March 4, 1997), p. A17.
36. Kathy Walt, "Group Pushes School Voucher Program," *Houston Chronicle* (March 4, 1997), p. A17.
37. "Fighting Vouchers in Texas: The 1999 Legislative Battle." Available at: http://www.tfn.org/issues/vouchers/1999legfight.htm.
38. The loan to Perry was also guaranteed by James Mansour and Houston investor William McMinn; Kim Cobb, "Texas a Big Voucher Battleground," *Houston Chronicle* (January 31, 1999), p. A1; "Fighting Vouchers in Texas: The 1999 Legislative Battle." Available at: http://www.tfn.org/issues/vouchers/1999legfight.htm.
39. Kathy Walt, "Group Pushes School Voucher Program," *Houston Chronicle* (March 4, 1997), p. A17.
40. A. Phillips Brooks, "Vouchers Opposed by Many Minority Lawmakers," *Austin American-Statesman* (June 6, 1996), pp. B1, B6.
41. A. Phillips Brooks, "Vouchers Opposed by Many Minority Lawmakers," *Austin American-Statesman* (June 6, 1996), p. B6.
42. A. Phillips Brooks, "Vouchers Opposed by Many Minority Lawmakers," *Austin American-Statesman* (June 6, 1996), p. B6.
43. Kent Grusendorf, "School Choice Deserves a Chance," *Austin American-Statesman* (March 30, 1994), p. A13.
44. House Research Organization, *Charter Schools, Vouchers, and Other School Choice Options*, HRO Publication No. 189 (Austin, June 28, 1994).
45. R. G. Ratcliffe, "Conservative Liberal with his Offerings," *Houston Chronicle* (September 21, 1997), pp. A1, A21.
46. R. G. Ratcliffe, "Conservative Liberal with his Offerings," *Houston Chronicle* (September 21, 1997), p. A21.
47. R. G. Ratcliffe, "Conservative Liberal with his Offerings," *Houston Chronicle* (September 21, 1997), p. A21.
48. R. G. Ratcliffe, "Conservative Liberal with his Offerings," *Houston Chronicle* (September 21, 1997), p. A21.
49. CPS is "a statewide organization that opposes vouchers but supports choice programs in public schools. The coalition includes the Texas State Teachers Association, the Texas Association of School Administrators and the Texas Parent Teachers Association, among others"; A. Phillips Brooks, "Texans Support School Vouchers," *Austin American-Statesman* (March 2, 1997), p. B1.

50. The letterhead on CPS stationary is emblazoned with the phrase "Private school vouchers: an experiment taxpayers can't afford"; A. Phillips Brooks, "Voucher Plans Open New Battle for Schools," *Austin American-Statesman* (March 13, 1997), p. B1.
51. A. Phillips Brooks, "Voucher Opponents Say They Can Block Bill in Senate," *Austin American-Statesman* (March 26, 1997), pp. B1, B9.
52. A. Phillips Brooks, "Texans Support School Vouchers," *Austin American-Statesman* (March 2, 1997), p. B1.
53. *ATPE Political Action Kit*, 1998.
54. Texas Classroom Teachers Association, "TCTA Legislative Program '97," (1998), available: www.tcta.org/leg97.htm; A. Phillips Brooks, "Delco: Texas Must Revitalize its Schools," *Austin American-Statesman* (March 24, 1994), p. B2.
55. A. Phillips Brooks, "Choice Programs Won't Fix Public Schools, Study Says," *Austin American-Statesman* (March 30, 1994), p. B2.
56. A. Phillips Brooks, "Choice Programs Won't Fix Public Schools, Study Says," *Austin American-Statesman* (March 30, 1994), p. B2.
57. A. Phillips Brooks, "Hispanic Groups to Grade Lawmakers," *Austin American-Statesman* (April 16, 1997), p. B3.
58. A. Phillips Brooks, "Limitations Abound in Education Vouchers," *Austin American-Statesman* (April 30, 1995), p. A15.
59. A. Phillips Brooks, "Limitations Abound in Education Vouchers," *Austin American-Statesman* (April 30, 1995), p. A15.
60. One proposal under consideration by the Senate during the 1995 legislative session would have provided scholarships to educationally disadvantaged students. In the letter, Gallegos stated that under the proposal, about 170,000 students would be eligible for the scholarships. This, he noted, is "170 times larger than the only other 'pilot' voucher program in the country which is in Milwaukee and larger than the entire student populations in 10 states." These scholarships could be used at private schools.
61. A. Phillips Brooks, "Delco: Texas Must Revitalize its Schools," *Austin American-Statesman* (March 24, 1994), p. B2.
62. A. Phillips Brooks, "House Votes Down Vouchers, Adopts Values Measures," *Austin American-Statesman* (May 7, 1995), p. B2.
63. A. Phillips Brooks, "House Approval Sends Education Bill to Negotiating Table," *Austin American-Statesman* (May 8, 1995), p. B3.
64. A. Phillips Brooks, "House Approval Sends Education Bill to Negotiating Table," *Austin American-Statesman* (May 8, 1995), p. B3.
65. A. Phillips Brooks, "Panel Supports School Vouchers," *Austin American-Statesman* (April 17, 1997), p. B8.
66. "Vouchers are No Cure," *Austin American-Statesman* (March 27, 1997), p. A14.
67. "Vouchers are No Cure," *Austin American-Statesman* (March 27, 1997), p. A14.

68. "Oppose School Vouchers," *Austin American-Statesman* (June 10, 1996), p. A8.
69. "Oppose School Vouchers," *Austin American-Statesman* (June 10, 1996), p. A8.
70. "Oppose School Vouchers," *Austin American-Statesman* (June 10, 1996), p. A8.
71. "Oppose School Vouchers," *Austin American-Statesman* (June 10, 1996), p. A8.
72. The pilot program that drew the most attention and had the best chance of passage called for 1,000 vouchers to be made available to children from low-income families. Currently, there are over 3.7 million students enrolled in public schools in Texas; A. Phillips Brooks, "Panel Supports School Vouchers," *Austin American-Statesman* (April 17, 1997), p. B8.
73. C. Richards, "Far Right Took Deserved Trouncing in '97 Legislature," *Austin American-Statesman* (June 17, 1997), p. A13.
74. C. Richards, "Far Right Took Deserved Trouncing in '97 Legislature," *Austin American-Statesman* (June 17, 1997), p. A13.
75. A. Phillips Brooks, "Vouchers Opposed by Many Minority Lawmakers," *Austin American-Statesman* (June 6, 1996), p. B6.
76. A. Phillips Brooks, "Voucher Plans Open New Battle for Schools," *Austin American-Statesman* (March 13, 1997), p. B5.
77. Ibid.; Cuellar and Wilson are by far the most vocal proponents of vouchers among minority legislators. In each of the last three legislative sessions, they have introduced voucher bills. In three record votes on private school vouchers during the 1997 legislative session, Wilson voted in favor of vouchers all three times, while Cuellar voted in favor of vouchers twice. They were the only minority members of the House to actively support vouchers.
78. *CER Newswire* 2:27 (July 12, 2000). Available at: http://www.edreform.com.
79. Jeff Archer, "NEA Dips into New Fund to Aid Campaigns," *Education Week* 20:4 (September 27, 2000), p. 17.
80. Tim L. Mazzoni, "State Policymaking and School Reform: Influences and Influentials," in Jay D. Scribner and Donald H. Layton, eds., *The Study of Educational Politics* (Washington, D.C.: Falmer Press, 1995), p. 67.
81. R. G. Ratcliffe, "Conservative Liberal with his Offerings," *Houston Chronicle* (September 21, 1997), pp. A1, A21.
82. R. G. Ratcliffe, "Conservative Liberal with his Offerings," *Houston Chronicle* (September 21, 1997), pp. A1, A21.
83. R. G. Ratcliffe, "Conservative Liberal with his Offerings," *Houston Chronicle* (September 21, 1997), pp. A1, A21.
84. This interpretation was supported by several members of both pro- and anti-voucher groups, as well as key legislative committee staff members and legislators.

85. Ken Herman, "Bullock Quits Voucher Group Chairmanship," *Austin American-Statesman* (March 7, 1998), pp. B1, B9.
86. Ken Herman, "Bullock Quits Voucher Group Chairmanship," *Austin American-Statesman* (March 7, 1998), pp. B1, B9.
87. Ken Herman, "Bullock Quits Voucher Group Chairmanship," *Austin American-Statesman* (March 7, 1998), pp. B1, B9.
88. Ken Herman, "Bullock Quits Voucher Group Chairmanship," *Austin American-Statesman* (March 7, 1998), pp. B1, B9.
89. The Texas Hispanic Families Coalition is composed of the League of United Latin American Citizens (LULAC), business, veterans and civil rights groups, including the Mexican American Legal Defense and Educational Fund (MALDEF) and the Texas Association of Mexican American Chambers of Commerce (Brooks, 1997d). This coalition supports charter schools but is opposed to vouchers.
90. CPS Position Paper, *The Choice Issue: Post Senate Bill 1 and Beyond* (Austin, no date).
91. "Fighting Vouchers in Texas: The 1999 Legislative Battle." Available at: http://www.tfn.org/issues/vouchers/1999legfight.htm.
92. "Fighting Vouchers in Texas: The 1999 Legislative Battle." Available at: http://www.tfn.org/issues/vouchers/1999legfight.htm.
93. "Fighting Vouchers in Texas: The 1999 Legislative Battle." Available at: http://www.tfn.org/issues/vouchers/1999legfight.htm.
94. Such tactics were incredibly damaging to the voucher movement in Texas and reflect the triumph of blind ideology over pragmatic policy-making. First, the pro-voucher incumbents would have supported voucher bills, though perhaps not the type of wide open voucher plan favored by the pro-voucher coalition. Second, in state and national politics, incumbents usually win (reelection rates vary but average approximately 90 percent). From the standpoint of practical politics, it is unwise to bet against or antagonize incumbents.
95. A. Phillips Brooks, "Delco: Texas Must Revitalize its Schools," *Austin American-Statesman* (March 24, 1994), pp. B1–B2.
96. A. Phillips Brooks, "Delco: Texas Must Revitalize its Schools," *Austin American-Statesman* (March 24, 1994), p. B2.
97. A. Phillips Brooks, "Delco: Texas Must Revitalize its Schools," *Austin American-Statesman* (March 24, 1994), p. B2.
98. A. Phillips Brooks, "Education Reform Unites Opponents," *Austin American-Statesman* (May 8, 1994), pp. B1, B7.
99. A. Phillips Brooks, "Sadler Says Special Session on Education not Needed," *Austin American-Statesman* (April 27, 1995), p. B3.
100. Texas Classroom Teachers Association, "TCTA Legislative Program '97," (1998), Available: www.tcta.org/leg97.htm.
101. A. Phillips Brooks, "Texas Doing Homework on Charter Schools Concept," *Austin American-Statesman* (February 20, 1994), p. A17.

102. A. Phillips Brooks, "Education Reform Unites Opponents," *Austin American-Statesman* (May 8, 1994), p. B7.
103. A. Phillips Brooks, "Bush Election Revives Controversial Initiatives," *Austin American-Statesman* (November 11, 1994), p. A8.
104. Dave McNeely, "Home-rule not Born of Racism," *Austin American-Statesman* (May 25, 1995), p. A11.
105. Paul A. Sabatier and Hank C. Jenkins-Smith, eds., *Policy Change and Learning: An Advocacy Coalition Approach* (Boulder, CO: Westview Press, 1993).
106. Competition restricted to within the public sector.
107. At least in theory, although some recent research questions this premise—asserting that charter school performance, evaluation, and renewal is as much a political issue as a technical "did students meet targeted performance goals?" issue. See Lance D. Fusarelli, "The Political Construction of Accountability," *Education and Urban Society* 33:2, pp. 157–169; Rick Hess, "Whaddya Mean You Want to Close My School?" *Education and Urban Society* 33:2, pp. 141–156.
108. Bryan C. Hassel, *The Charter School Challenge* (Washington, D.C.: Brookings Institution Press, 1999), p. 2.
109. This coalition supports quality education initiatives for African Americans.
110. This is a coalition of major employers and businesses working to improve Texas's public schools.
111. Joseph P. Viteritti, "Coming Around on School Choice," *Educational Leadership* 59:7 (April 2002), p. 45.
112. Paul C. Bauman, *Governing Education: Public Sector Reform or Privatization* (Boston: Allyn and Bacon, 1996); Chris Pipho, "Bipartisan Charter Schools," *Phi Delta Kappan* 75:3 (1993), pp. 102–103.
113. C. Clark, "The Growing Movement Toward Charter Schools," *Congressional Quarterly Researcher* 6:28 (1996), p. 656; Priscilla Wohlstetter and Lesley Anderson, "What Can U.S. Charter Schools Learn from England's Grant-Maintained Schools?" *Phi Delta Kappan* 75:6 (February 1994), pp. 486–491.
114. C. Clark, "The Growing Movement Toward Charter Schools," *Congressional Quarterly Researcher* 6:28 (1996), p. 656.
115. Organizations such as the Center for Education Reform (a school choice advocacy group) differ considerably from organizations such as the American Federation of Teachers in their definition of what constitutes "strong" and "weak" charter school statutes (what one group considers a strong attribute, the other considers weak). Following Hassel, we use what is becoming the generally accepted definition of a "strong" charter school statute—one with multiple approval bodies, low entry barriers, few restrictions on applicants/organizers, legal and fiscal independence, substantial regulatory exemptions, and no caps on the number of charters that can be granted. See Bryan C. Hassel,

The Charter School Challenge (Washington, D.C.: Brookings Institution Press, 1999), pp. 18–19.

116. In the deal, state lawmakers received a 38 percent pay raise. In addition, "Statewide elected officials, judges, and commissioners also received pay hikes, and the salary of the governor rose from $130,000 to $179,000," Sandra Vergari, "New York: Over 100 Charter Applications in Year One," in Sandra Vergari, ed., *The Charter School Landscape* (Pittsburgh: University of Pittsburgh Press, 2002), p. 285; Clifford J. Levy and Anemona Hartocollis, "Crew Assails Albany Accord on Opening Charter Schools," *New York Times* (December 19, 1998), pp. D1, D2.
117. Chris Pipho, "Bipartisan Charter Schools," *Phi Delta Kappan* 75:3 (1993), pp. 102–103.
118. Chris Pipho, "Bipartisan Charter Schools," *Phi Delta Kappan* 75:3 (1993), pp. 102–103.
119. Eric Hirsch, "Colorado Charter Schools: Becoming an Enduring Feature of the Reform Landscape," in Sandra Vergari, ed., *The Charter School Landscape* ((Pittsburgh: University of Pittsburgh Press, 2002), pp. 93–112.
120. Eric Hirsch, "Colorado Charter Schools: Becoming an Enduring Feature of the Reform Landscape," in Sandra Vergari, ed., *The Charter School Landscape* ((Pittsburgh: University of Pittsburgh Press, 2002), p. 94.
121. The OEA, representing over 114,000 members, is the state's largest teachers' union; "Charter Schools Proposal Clears Hurdle," *The Plain Dealer* (March 26, 1996), p. B2.
122. Mary Beth Lane, "Voinovich Plans to Mediate Charter Schools Legislation," *The Plain Dealer* (February 28, 1996), p. B5.
123. Mary Beth Lane, "Voinovich Plans to Mediate Charter Schools Legislation," *The Plain Dealer* (February 28, 1996), p. B5.
124. Mary Beth Lane, "Experiments in Education with Two Bills in Legislature, It's likely Ohio Will Try Community Schools," *The Plain Dealer* (July 25, 1995), p. B4.
125. "All Chatter; No Charters," *The Plain Dealer* (June 16, 1996), p. C2.
126. Chris Sheridan, "Horse-Trading Ahead on Schools," *The Plain Dealer* (November 10, 1996), p. H2.
127. Scott Stephens and Alison Grant, "Teachers Open to Charter Schools, Support Depends on How Law is Written," *The Plain Dealer* (November 12, 1996), p. B1.
128. Scott Stephens and Alison Grant, "Teachers Open to Charter Schools, Support Depends on How Law is Written," *The Plain Dealer* (November 12, 1996), p. B1.
129. Scott Stephens and Alison Grant, "Teachers Open to Charter Schools, Support Depends on How Law is Written," *The Plain Dealer* (November 12, 1996), p. B1.
130. "Charter Schools Proposal Clears Hurdle," *The Plain Dealer* (March 26, 1996), p. B2.

131. Peter J. Shelly, "Charter School Bill Advances," *Pittsburgh Post-Gazette* (March 14, 1996), p. A14.
132. Frank Reeves, "30-18 Vote Send Charter Schools to House," *Pittsburgh Post-Gazette* (June 12, 1997), p. A13.
133. Frank Reeves, "30-18 Vote Send Charter Schools to House," *Pittsburgh Post-Gazette* (June 12, 1997), p. A13.
134. D. A. Taebel and C. T. Brenner, eds., *Privatizing Public Education: The Texas Experience* (Arlington, TX: Institute of Urban Studies & Center for Economic Development Research and Service, 1994), p. 99.
135. D. A. Taebel and C. T. Brenner, eds., *Privatizing Public Education: The Texas Experience* (Arlington, TX: Institute of Urban Studies & Center for Economic Development Research and Service, 1994), p. 99.
136. A. Phillips Brooks, "Choice Programs Won't Fix Public Schools, Study Says," *Austin American-Statesman* (March 30, 1994), pp. B1–B2.
137. House Research Organization, *Charter Schools, Vouchers, and Other School Choice Options*, HRO Publication No. 189 (Austin, June 28, 1994).
138. Robert C. Bulman and David L. Kirp, "The Shifting Politics of School Choice," in Stephen D. Sugarman and Frank R. Kemerer, eds., *School Choice and Social Controversy* (Washington, D.C.: Brookings Institution Press, 1999), pp. 36–67; C. R. Sauter, *Charter Schools: A New Breed of Public Schools* (Oak Brook, IL: The North Central Regional Educational Laboratory, 1993).
139. Robert C. Bulman and David L. Kirp, "The Shifting Politics of School Choice," in Stephen D. Sugarman and Frank R. Kemerer, eds., *School Choice and Social Controversy* (Washington, D.C.: Brookings Institution Press, 1999), p. 52.
140. Bella Rosenberg of the American Federation of Teachers argues that some charter schools clearly represent an attempt to establish private schools within the public school system. One could argue that the distinction made between the public and private spheres is arbitrary, largely rhetorical, and socially constructed. As such, the distinction between charter schools and voucher plans is far less than the rhetoric of policy debates (and the conflict generated therein) would suggest. This will become clear should a voucher system ever be implemented in Texas. The rules and regulations governing the program (state mandates) may be similar to those that currently govern charter schools.
141. K. McGree, *Redefining Education Governance: The Charter School Concept* (Austin: Southwest Educational Development Laboratory, 1995), p. 11.
142. Joseph P. Viteritti, *Choosing Equality: School Choice, the Constitution, and Civil Society* (Washington, D.C.: The Brookings Institution, 1999), p. 62.
143. Robert C. Bulman and David L. Kirp, "The Shifting Politics of School Choice," in Stephen D. Sugarman and Frank R. Kemerer, eds., *School*

Choice and Social Controversy (Washington, D.C.: Brookings Institution Press, 1999), pp. 40–43.

144. Michelle Godard McNiff and Bryan C. Hassel, "Charter Schools in North Carolina: Confronting the Challenges of Rapid Growth," in Sandra Vergari, ed., *The Charter School Landscape* (Pittsburgh: University of Pittsburgh Press, 2002), p. 209.
145. Frederick M. Hess and Robert Maranto, "Letting a Thousand Flowers (and Weeds) Bloom: The Charter Story in Arizona," in Sandra Vergari, ed., *The Charter School Landscape* (Pittsburgh: University of Pittsburgh Press, 2002), p. 59.
146. Frederick M. Hess and Robert Maranto, "Letting a Thousand Flowers (and Weeds) Bloom: The Charter Story in Arizona," in Sandra Vergari, ed., *The Charter School Landscape* (Pittsburgh: University of Pittsburgh Press, 2002), p. 59.
147. Robert C. Bulman and David L. Kirp, "The Shifting Politics of School Choice," in Stephen D. Sugarman and Frank R. Kemerer, eds., *School Choice and Social Controversy* (Washington, D.C.: Brookings Institution Press, 1999), p. 54.
148. Priscilla Wohlstetter, Noelle C. Griffin, and Derrick Chau, "Charter Schools in California: A Bruising Campaign for Public School Choice," in Sandra Vergari, ed., *The Charter School Landscape* (Pittsburgh: University of Pittsburgh Press, 2002), pp. 32–53.
149. Charles Mahtesian, "Teachers Union Taught a Lesson in Political Arena," *The Plain Dealer* (December 14, 1995), p. A28.
150. Priscilla Wohlstetter, Noelle C. Griffin, and Derrick Chau, "Charter Schools in California: A Bruising Campaign for Public School Choice," in Sandra Vergari, ed., *The Charter School Landscape* (Pittsburgh: University of Pittsburgh Press, 2002), pp. 32–53.
151. Robert C. Bulman and David L. Kirp, "The Shifting Politics of School Choice," in Stephen D. Sugarman and Frank R. Kemerer, eds., *School Choice and Social Controversy* (Washington, D.C.: Brookings Institution Press, 1999), p. 66.
152. Bruce S. Cooper and E. Vance Randall, *Critics Stop Vouchers in Their Tracks.* paper presented at the Education Between State and Civil Society Conference. Boston, MA, 1997, pp. 3–4.
153. Bruce S. Cooper and E. Vance Randall, *Critics Stop Vouchers in Their Tracks.* paper presented at the Education Between State and Civil Society Conference. Boston, MA, 1997, p. 4.
154. Robert C. Bulman and David L. Kirp, "The Shifting Politics of School Choice," in Stephen D. Sugarman and Frank R. Kemerer, eds., *School Choice and Social Controversy* (Washington, D.C.: Brookings Institution Press, 1999), p. 61.
155. Anemona Hartocollis, "Giving up Attack, Crew Offers his Charter School Plan," *New York Times* (February 2, 1999), p. B4.
156. Anemona Hartocollis, "Giving up Attack, Crew Offers his Charter School Plan," *New York Times* (February 2, 1999), p. B4

157. Joseph P. Viteritti, *Choosing Equality: School Choice, the Constitution, and Civil Society* (Washington, D.C.: The Brookings Institution, 1999), p. 70.
158. Chester E. Finn, Jr., Bruno V. Manno, and Gregg Vanourek, *Charter Schools in Action* (Princeton: Princeton University Press, 2000), p. 178.
159. Chester E. Finn, Jr., Bruno V. Manno, and Gregg Vanourek, *Charter Schools in Action* (Princeton: Princeton University Press, 2000), p. 183.
160. *CER Newswire* 3:15 (April 10, 2001). Available at: http://www.edreform.com.
161. Chester E. Finn, Jr., Bruno V. Manno, and Gregg Vanourek, *Charter Schools in Action* (Princeton: Princeton University Press, 2000), p. 180; Frederick M. Hess and Robert Maranto, "Letting a Thousand Flowers (and Weeds) Bloom: The Charter Story in Arizona," in Sandra Vergari, ed., *The Charter School Landscape* (Pittsburgh: University of Pittsburgh Press, 2002), pp. 54–73.
162. Robert C. Bulman and David L. Kirp, "The Shifting Politics of School Choice," in Stephen D. Sugarman and Frank R. Kemerer, eds., *School Choice and Social Controversy* (Washington, D.C.: Brookings Institution Press, 1999), p. 56.
163. Michael Mintrom and David N. Plank, "School Choice in Michigan," in Paul E. Peterson and David E. Campbell, eds., *Charters, Vouchers, and Public Education* (Washington, D.C.: Brookings Institution Press, 2001), pp. 43–58.
164. Mary Ann Zehr, "Charters in Some Cities Attract Students from Catholic Schools," *Education Week* 21:37 (May 22, 2002), p. 12.
165. Mary Ann Zehr, "Charters in Some Cities Attract Students from Catholic Schools," *Education Week* 21:37 (May 22, 2002), p. 12.
166. Michael Mintrom and David N. Plank, "School Choice in Michigan," in Paul E. Peterson and David E. Campbell, eds., *Charters, Vouchers, and Public Education* (Washington, D.C.: Brookings Institution Press, 2001), p. 58.
167. *CER Newswire* 3:19 (May 8, 2001). Available at: http://www.edreform.com.
168. *CER Newswire* 3:13 (March 27, 2001). Available at: http://www.edreform.com.
169. *CER Newswire* 4:20 (May 28, 2002). Available at: http://www.edreform.com.
170. Robert C. Bulman and David L. Kirp, "The Shifting Politics of School Choice," in Stephen D. Sugarman and Frank R. Kemerer, eds., *School Choice and Social Controversy* (Washington, D.C.: Brookings Institution Press, 1999), pp. 54–55.
171. *CER Newswire* 3:19 (May 8, 2001). Available at: http://www.edreform.com.
172. Chester E. Finn, Jr., Bruno V. Manno, and Gregg Vanourek, *Charter Schools in Action* (Princeton: Princeton University Press, 2000).

173. *CER Newswire* 4:20 (May 28, 2002). Available at: http://www.edreform.com.
174. Cited in James G. Cibulka, "The Reform and Survival of American Public Schools: An Institutional Perspective," in Robert L. Crowson, William L. Boyd, and Hanne B. Mawhinney, eds., *The Politics of Education and the New Institutionalism* (London: Falmer Press, 1996), pp. 7–22.
175. J. Brooke, "Denver Minorities Sue for School Vouchers," *Austin American-Statesman* (December 27, 1997), p. A5.
176. Michael Mintrom and David N. Plank, "School Choice in Michigan," in Paul E. Peterson and David E. Campbell, eds., *Charters, Vouchers, and Public Education* (Washington, D.C.: Brookings Institution Press, 2001), pp. 43–58.
177. Recall that an early voucher plan—the federally sponsored voucher program in Alum Rock, California, was originally conceived by liberal Democratic policy works in the Johnson Administration who viewed it as another tool of redistributive social policy.
178. T. Toch, "Liberal Vouching for City Schools," *U.S. News & World Report* 122:23 (1997), p. 40.
179. Jim Carl, "Unusual Allies: Elite and Grass-Roots Origins of Parental Choice in Milwaukee," *Teachers College Record* 98:2 (1996), p. 279.
180. Joseph P. Viteritti, "Coming Around on School Choice," *Educational Leadership* 59:7 (April 2002), p. 47.
181. Ibid.; Viteritti notes that in the 2001 mayoral election in New York City, only one of the six candidates supported vouchers for poor children, yet all sent their own children to private schools. Furthermore, the New York City schools chancellor, as well as nearly the entire New York City board of education, the mayor and former mayor, governor and former governor, and newly elected U.S. senator all opposed vouchers for poor children, while "refusing to send their own children to public schools" (p. 47).
182. Carl Campanile, "Cardinal in Holy War Over Schools," *New York Post* (September 2, 2002), p. A1.
183. Jim Carl, "Unusual Allies: Elite and Grass-Roots Origins of Parental Choice in Milwaukee," *Teachers College Record* 98:2 (1996), p. 273.
184. Robert C. Bulman and David L. Kirp, "The Shifting Politics of School Choice," in Stephen D. Sugarman and Frank R. Kemerer, eds., *School Choice and Social Controversy* (Washington, D.C.: Brookings Institution Press, 1999), p. 49.
185. J. J. Miller, "Why School Choice Lost," *The Wall Street Journal* (November 4, 1993), p. A8.
186. Hispanic and African American legislators were largely silent or quiescent on the issue of charter schools. The lack of organized opposition facilitated the passage of this legislation.
187. Scott Stephens, "Storming the Statehouse, Local Voucher Plan Supporters State their Case in Columbus," *The Plain Dealer* (February 1, 1995), p. B1.

188. Scott Stephens, "Storming the Statehouse, Local Voucher Plan Supporters State their Case in Columbus," *The Plain Dealer* (February 1, 1995), p. B1.
189. Ben E. Espy, " 'Separate but Equal' Public Funding for Vouchers and Charter Schools Will Create Academic Segregation in Ohio," *The Plain Dealer* (June 17, 1996), p. B9.
190. James Lawless, "School Officials Oppose Voucher Plan," *The Plain Dealer* (March 4, 1995), p. B1.
191. Brennan is profiled in Hubert Morken and Jo Renee Formicola's book *The Politics of School Choice* (Lanham, MD: Rowman & Littlefield, 1999), pp. 261–262; James Lawless, "School Officials Oppose Voucher Plan," *The Plain Dealer* (March 4, 1995), p. B1.
192. Fances C. Fowler, "Education Reform Comes to Ohio: An Application of Mazzoni's Arena Models," *Educational Evaluation and Policy Analysis* 16:3, pp. 335–350.
193. Joe Hallett, "Group Asks Legislators to End Voucher Program," *The Plain Dealer* (May 14, 1997), p. A1.
194. James Lawless, "School Officials Oppose Voucher Plan," *The Plain Dealer* (March 4, 1995), p. B1.
195. Peter J. Shelly and Frank Reeves, "Ridge Shelves Voucher Plan," *Pittsburgh Post-Gazette* (December 14, 1995), p. A1.
196. Peter J. Shelly and Frank Reeves, "Common Ground Sought on Vouchers," *Pittsburgh Post-Gazette* (June 6, 1999), p. C5.
197. Hubert Morken and Jo Renee Formicola, *The Politics of School Choice* (Lanham, MD: Rowman & Littlefield, 1999).
198. David Boldt, "School Choice and Anti-Catholicism," *Pittsburgh Post-Gazette* (June 16, 1999), p. A2.
199. Peter J. Shelly and Frank Reeves, "Ridge Shelves Voucher Plan," *Pittsburgh Post-Gazette* (December 14, 1995), p. A1.
200. Peter J. Shelly and Frank Reeves, "Bell Rings for Round 2 of School Vouchers," *Pittsburgh Post-Gazette* (November 19, 1995), p. A19.
201. Peter J. Shelly and Frank Reeves, "Bell Rings for Round 2 of School Vouchers," *Pittsburgh Post-Gazette* (November 19, 1995), p. A19.
202. Peter J. Shelly and Frank Reeves, "Bell Rings for Round 2 of School Vouchers," *Pittsburgh Post-Gazette* (November 19, 1995), p. A19.
203. Hubert Morken and Jo Renee Formicola, *The Politics of School Choice* (Lanham, MD: Rowman & Littlefield, 1999).
204. Hubert Morken and Jo Renee Formicola, The Politics of School Choice (Lanham, MD: Rowman & Littlefield, 1999).
205. Hubert Morken and Jo Renee Formicola, The Politics of School Choice (Lanham, MD: Rowman & Littlefield, 1999), p. 74.
206. Robert C. Bulman and David L. Kirp, "The Shifting Politics of School Choice," in Stephen D. Sugarman and Frank R. Kemerer, eds., *School Choice and Social Controversy* (Washington, D.C.: Brookings Institution Press, 1999), pp. 36–67.

207. REACH (The Road to Educational Achievement through Choice) is Pennsylvania's school choice coalition.
208. Frank Reeves and Peter J. Shelly, "Day Care Fight Imperils Voucher Plan," *Pittsburgh Post-Gazette* (April 14, 1999), p. B1.
209. Frank Reeves and Peter J. Shelly, "Day Care Fight Imperils Voucher Plan," *Pittsburgh Post-Gazette* (April 14, 1999), p. B1.
210. Peter J. Shelly and Frank Reeves, "Ridge Open to Testing of Voucher Students," *Pittsburgh Post-Gazette* (June 4, 1999), p. B1.
211. *CAPE Update* (November 8, 2000). Available at: http://www.capenet.org.
212. *CAPE Update* (November 8, 2000). Available at: http://www.capenet.org.
213. *CAPE Update* (November 8, 2000). Available at: http://www.capenet.org.
214. Carolyn Hess, "No Choice in Rural Areas," *Pittsburgh Post-Gazette* (June 14, 1999), p. A14.
215. Michael Mintrom and David N. Plank, "School Choice in Michigan," in Paul E. Peterson and David E. Campbell. eds., *Charters, Vouchers, and Public Education* (Washington, D.C.: Brookings Institution Press, 2001), p. 46.
216. Michael Mintrom and David N. Plank, "School Choice in Michigan," in Paul E. Peterson and David E. Campbell, eds., *Charters, Vouchers, and Public Education* (Washington, D.C.: Brookings Institution Press, 2001), p. 50.
217. Hubert Morken and Jo Renee Formicola, *The Politics of School Choice* (Lanham, MD: Rowman & Littlefield, 1999).
218. Hubert Morken and Jo Renee Formicola, *The Politics of School Choice* (Lanham, MD: Rowman & Littlefield, 1999), pp. 95–96.
219. Frank Reeves and Peter J. Shelly, "Day Care Fight Imperils Voucher Plan," *Pittsburgh Post-Gazette* (April 14, 1999), p. B1.
220. Frank Reeves and Peter J. Shelly, "Day Care Fight Imperils Voucher Plan," *Pittsburgh Post-Gazette* (April 14, 1999), p. B1.
221. Frank Reeves and Peter J. Shelly, "Day Care Fight Imperils Voucher Plan," *Pittsburgh Post-Gazette* (April 14, 1999), p. B1.
222. Frank Reeves and Peter J. Shelly, "Day Care Fight Imperils Voucher Plan," *Pittsburgh Post-Gazette* (April 14, 1999), p. B1.
223. Florida's statewide voucher bill, signed into law by Governor Jeb Bush, requires voucher recipients to return to their home district for testing; Peter J. Shelly and Frank Reeves, "Ridge Open to Testing of Voucher Students," *Pittsburgh Post-Gazette* (June 4, 1999), p. B1; Frank Reeves, "Critics Cite Private School Testing," *Pittsburgh Post-Gazette* (June 13, 1999), p. B7.
224. Peter J. Shelly and Frank Reeves, "Ridge Open to Testing of Voucher Students," *Pittsburgh Post-Gazette* (June 4, 1999), p. B1.
225. Frank Reeves, "Rep. Preston Says Remarks on Vouchers Were Misconstrued," *Pittsburgh Post-Gazette* (June 11, 1999), p. B1.

226. Frank Reeves, "Rep. Preston Says Remarks on Vouchers Were Misconstrued," *Pittsburgh Post-Gazette* (June 11, 1999), p. B1.
227. David Boldt, "School Choice and Anti-Catholicism," *Pittsburgh Post-Gazette* (June 16, 1999), p. A2.
228. Michael F. Holt, *Political Parties and American Political Development* (Baton Rouge, LA: Louisiana State University Press, 1992), p. 89.
229. Michael F. Holt, *Political Parties and American Political Development* (Baton Rouge, LA: Louisiana State University Press, 1992), p. 89.
230. Michael F. Holt, *Political Parties and American Political Development* (Baton Rouge, LA: Louisiana State University Press, 1992), p. 118.
231. Approximately one-third of the population in Pennsylvania identifies themselves as Catholic. (See John J. Kennedy, *The Contemporary Pennsylvania Legislature* [Lanham, MD: University Press of America, 1999], p. 20.)
232. Cited in Rick Hinshaw, "Anti-Catholicism Today," in Robert P. Lockwood, ed., *Anti-Catholicism in American Culture* (Huntington, IN: Our Sunday Visitor, Inc., 2000), p. 89.
233. Robert P. Lockwood, "Introduction," in Robert P. Lockwood, ed., *Anti-Catholicism in American Culture* (Huntington, IN: Our Sunday Visitor, Inc., 2000), p. 5.
234. Andrew M. Greeley, *An Ugly Little Secret: Anti-Catholicism in North America* (Kansas City, MO: Sheed Andrews & McNeel, Inc., 1977), p. 1.
235. Robert P. Lockwood, "The Evolution of Anti-Catholicism in the United States," in Robert P. Lockwood, ed., *Anti-Catholicism in American Culture* (Huntington, IN: Our Sunday Visitor, Inc., 2000), pp. 15–53.
236. Andrew M. Greeley, *An Ugly Little Secret: Anti-Catholicism in North America* (Kansas City, MO: Sheed Andrews & McNeel, Inc., 1977), p. 79.
237. Joseph P. Viteritti, *Choosing Equality: School Choice, the Constitution, and Civil Society* (Washington, D.C.: Brookings Institution Press, 1999), p. 10.
238. Readers interested in an extended, historical discussion of the role of anti-Catholic sentiment in school choice should consult Joseph P. Viteritti, *Choosing Equality: School Choice, the Constitution, and Civil Society* (Washington, D.C.: Brookings Institution Press, 1999), pp. 146–155.
239. Andrew M. Greeley, *An Ugly Little Secret: Anti-Catholicism in North America* (Kansas City, MO: Sheed Andrews & McNeel, Inc., 1977), p. 81.
240. Peter J. Shelly and Frank Reeves, "Bell Rings for Round 2 of School Vouchers," *Pittsburgh Post-Gazette* (November 19, 1995), p. A19.
241. Peter J. Shelly and Frank Reeves, "School Choice Decision Expected, Legislature Plans Three Days of Voting," *Pittsburgh Post-Gazette* (June 13, 1999), p. B1.

242. Peter J. Shelly and Frank Reeves, "School Choice Decision Expected, Legislature Plans Three Days of Voting," *Pittsburgh Post-Gazette* (June 13, 1999), p. B1.
243. Peter J. Shelly and Frank Reeves, "School Choice Decision Expected, Legislature Plans Three Days of Voting," *Pittsburgh Post-Gazette* (June 13, 1999), p. B1.
244. Frederick M. Hess and Robert Maranto, "Letting a Thousand Flowers (and Weeds) Bloom: The Charter Story in Arizona," in Sandra Vergari, ed., *The Charter School Landscape* (Pittsburgh: University of Pittsburgh Press, 2002), p. 56.
245. Michael Mintrom and Sandra Vergari, "Advocacy Coalitions, Policy Entrepreneurs, and Policy Change," *Policy Studies Journal* 24:3 (1996), pp. 420–434.
246. Bryan C. Hassel, *The Charter School Challenge* (Washington, D.C.: Brookings Institution Press, 1999), p. 48.
247. An affiliate of the National Education Association.
248. An affiliate of the American Federation of Teachers.
249. Bryan C. Hassel, *The Charter School Challenge* (Washington, D.C.: Brookings Institution Press, 1999).
250. Bryan C. Hassel, *The Charter School Challenge* (Washington, D.C.: Brookings Institution Press, 1999), p. 51.
251. Bryan C. Hassel, *The Charter School Challenge* (Washington, D.C.: Brookings Institution Press, 1999), p. 51.
252. Jane H. Karper and William Lowe Boyd, "Interest Groups and the Changing Environment of State Educational Policymaking: Developments in Pennsylvania," *Educational Administration Quarterly* 24:1, p. 28.
253. Jane H. Karper and William Lowe Boyd, "Interest Groups and the Changing Environment of State Educational Policymaking: Developments in Pennsylvania," *Educational Administration Quarterly* 24:1, pp. 21–54.
254. Jane H. Karper and William Lowe Boyd, "Interest Groups and the Changing Environment of State Educational Policymaking: Developments in Pennsylvania," *Educational Administration Quarterly* 24:1, p. 42.
255. Jane H. Karper and William Lowe Boyd, "Interest Groups and the Changing Environment of State Educational Policymaking: Developments in Pennsylvania," *Educational Administration Quarterly* 24:1, p. 44.

CHAPTER 5 ORGANIZATIONAL LEARNING DYNAMICS: THE UTILITY OF EXPERIENCE

1. Harald Baldersheim and Per Stava, "Reforming Local Government Policymaking and Management Through Organizational Learning and Experimentation: The Case of Norway," *Policy Studies Journal* 21:1 (1993), pp. 105–106.

2. Chris Argyris and Donald A. Schon, *Organizational Learning: A Theory of Action Perspective* (Reading, PA: Addison-Wesley, 1978), p. 19.
3. Frans L. Leeuw and Richard C. Sonnichsen, "Introduction: Evaluation and Organizational Learning: International Perspectives," in Frans L. Leeuw, Ray C. Rist, and Richard C. Sonnichsen, eds., *Can Governments Learn? Comparative Perspectives on Evaluation & Organizational Learning* (New Brunswick, NJ: Transaction Publishers, 1994), pp. 1–13.
4. Mariann Jelinek, *Institutionalizing Innovation: A Study of Organizational Learning Systems* (New York: Praeger, 1979), p. xii.
5. Mariann Jelinek, *Institutionalizing Innovation: A Study of Organizational Learning Systems* (New York: Praeger, 1979), p. xx.
6. Chris Argyris and Donald A. Schon, *Organizational Learning: A Theory of Action Perspective* (Reading, PA: Addison-Wesley, 1978), p. 20.
7. Harald Baldersheim and Per Stava, "Reforming Local Government Policymaking and Management Through Organizational Learning and Experimentation: The Case of Norway," *Policy Studies Journal* 21:1 (1993), p. 106.
8. Peter A. Hall, "Policy Paradigms, Social Learning, and the State: The Case of Economic Policymaking in Britain," *Comparative Politics* 25:3 (1993), p. 293.
9. Giandomenico Majone, *Evidence, Argument and Persuasion in the Policy Process* (New Haven: Yale University Press, 1989).
10. Hanne B. Mawhinney, "Towards an Ethical Policy Analysis: A Schema for the Interpretation of Argument and Debate in Educational Policy Change," paper presented at the Annual Meeting of the American Educational Research Association, New Orleans, LA (1994), p. 1.
11. Carol H. Weiss and M. J. Bucuvalas, "The Challenge of Social Research to Decisionmaking," in Carol H. Weiss, ed., *Using Social Research in Public Policymaking* (Lexington, MA: D.C. Heath and Company, 1977), p. 226.
12. Carol H. Weiss and M. J. Bucuvalas, "The Challenge of Social Research to Decisionmaking," in Carol H. Weiss ed., *Using Social Research in Public Policymaking* (Lexington, MA: D.C. Heath and Company, 1977), p. 226.
13. William N. Dunn, *Public Policy Analysis: An Introduction* (Englewood Cliffs, NJ: Prentice Hall, 1994).
14. L. J. Sharpe, "The Social Scientist and Policymaking: Some Cautionary Thoughts and Transatlantic Reflections," in Carol H. Weiss, ed., *Using Social Research in Public Policymaking* (Lexington, MA: D.C. Heath and Company, 1977), p. 45.
15. Daniel Callahan and Bruce Jennings, "Introduction," in Daniel Callahan and Bruce Jennings, eds., *Ethics, the Social Sciences and Policy Analysis* (New York: Plenum Press, 1983), p. xiii.
16. Mary E. Hawkesworth, *Theoretical Issues in Policy Analysis* (Albany: State University of New York Press, 1988), p. 28.

17. Daniel Callahan and Bruce Jennings, "Introduction," in Daniel Callahan and Bruce Jennings, eds., *Ethics, the Social Sciences and Policy Analysis* (New York: Plenum Press, 1983), p. xiii.
18. Carol H. Weiss, "Introduction," in Carol H. Weiss, ed., *Using Social Research in Public Policymaking* (Lexington, MA: D. C. Heath and Company, 1977), p. 10.
19. R. Mayntz, "Sociology, Value Freedom, and the Problems of Political Counseling," in Carol H. Weiss, ed., *Using Social Research in Public Policymaking* (Lexington, MA: D. C. Heath and Company, 1977), p. 62.
20. Paul A. Sabatier, "Toward Better Theories of the Policy Process," *PS: Political Science & Politics* 24:2 (1991), p. 148.
21. Carol H. Weiss, "Introduction," in Carol H. Weiss, ed., *Using Social Research in Public Policymaking* (Lexington, MA: D. C. Heath and Company, 1977), p. 14; see also Frank W. Lutz, "Policy-oriented Research: What Constitutes Good Proof," *Theory into Practice* 27:2 (1988), pp. 126–131.
22. Frank W. Lutz, "Policy-oriented Research: What Constitutes Good Proof," *Theory into Practice* 27:2 (1988), pp. 126–131.
23. Deborah A. Stone, *Policy Paradox and Political Reason* (Glenview, IL: Scott Foresman, 1988).
24. Paul A. Sabatier, "Toward Better Theories of the Policy Process," *PS: Political Science & Politics* 24:2 (1991), p. 148.
25. Carol H. Weiss, "Ideology, Interests, and Information: The Basis of Policy Decisions," in Daniel Callahan and Bruce Jennings, eds., *Ethics, the Social Sciences and Policy Analysis* (New York: Plenum Press, 1983), p. 219.
26. J. Mayne, "Utilizing Evaluation in Organizations: The Balancing Act," in Frans L. Leeuw, Ray C. Rist, and Richard C. Sonnichsen, eds., *Can Governments Learn? Comparative Perspectives on Evaluation & Organizational Learning* (New Brunswick, NJ: Transaction Publishers, 1994), p. 18.
27. Mariann Jelinek, *Institutionalizing Innovation: A Study of Organizational Learning Systems* (New York: Praeger, 1979), p. xii.
28. Notable exceptions include Chris Argyris and Donald A. Schon, *Organizational Learning: A Theory of Action Perspective* (Reading, PA: Addison-Wesley, 1978); Peter A. Hall, "Conclusion: The Political Power of Economic Ideas," in Peter A. Hall, ed., *The Political Power of Economic Ideas: Keynesianism Across Nations* (Princeton: Princeton University Press, 1989), pp. 361–392; Ray C. Rist, "The Preconditions for Learning: Lessons from the Public Sector," in Frans L. Leeuw, Ray C. Rist, and Richard C. Sonnichsen, eds., *Can Governments Learn? Comparative Perspectives on Evaluation and Organizational Learning* (New Brunswick, NJ: Transaction Publishers, 1994), pp. 189–205; Richard Rose, "What is Lesson-Drawing?" *Journal of Public Policy* 11 (1991), pp. 3–30; Richard Rose, *Lesson-Drawing in Public Policy* (Chatham, NJ: Chatham House, 1993); P. Shrivastava, "A Typology of

Organizational Learning Systems," *Journal of Management Studies* 20 (1983), pp. 7–28.

29. Peter A. Hall, "Conclusion: The Political Power of Economic Ideas," in Peter A. Hall, ed., *The Political Power of Economic Ideas: Keynesianism Across Nations* (Princeton: Princeton University Press, 1989), pp. 361–392.
30. Richard Rose, "What is Lesson-Drawing?" *Journal of Public Policy* 11 (1991), pp. 3–30; Richard Rose, *Lesson-Drawing in Public Policy* (Chatham, NJ: Chatham House, 1993).
31. Ray C. Rist, "The Preconditions for Learning: Lessons from the Public Sector," in Frans L. Leeuw, Ray C. Rist, and Richard C. Sonnichsen, eds., *Can Governments Learn? Comparative Perspectives on Evaluation and Organizational Learning* (New Brunswick, NJ: Transaction Publishers, 1994), p. 189.
32. Michael G. Fullan with S. Stiegelbauer, *The New Meaning of Educational Change*, 2nd ed. (Ontario: Ontario Institute for Studies in Education, 1991), p. 66.
33. Richard F. Elmore and Milbrey W. McLaughlin, *Steady Work: Policy Practice and the Reform of American Education*, Rand Publication Series No. R-3574-NIE/RC (Santa Monica, CA: National Institute of Education and Rand, 1988).
34. Mary E. Hawkesworth, *Theoretical Issues in Policy Analysis* (Albany: State University of New York Press, 1988); Hanne B. Mawhinney, "An Interpretive Framework for Understanding the Politics of Policy Change," paper presented at the Annual Meeting of the Canadian Association for Studies in Educational Administration, Calgary, Canada (1994).
35. Paul A. Sabatier and Hank C. Jenkins-Smith, eds., *Policy Change and Learning: An Advocacy Coalition Approach* (Boulder, CO: Westview Press, 1993).
36. Jerry L. Patterson, Stewart C. Purkey, and Jackson V. Parker, *Productive School Systems for a Nonrational World* (Alexandria, VA: Association for Supervision and Curriculum Development, 1986).
37. Available: www.ncsl.org/statevote98/tottrn.htm. Accessed October 20, 2002.
38. Typically, even states with low turnover experience turnover rates of 10–20 percent.
39. Frans L. Leeuw and Richard C. Sonnichsen, "Introduction: Evaluation and Organizational Learning: International Perspectives," in Frans L. Leeuw, Ray C. Rist, and Richard C. Sonnichsen, eds., *Can Governments Learn? Comparative Perspectives on Evaluation & Organizational Learning* (New Brunswick, NJ: Transaction Publishers, 1994), p. 1.
40. Frans L. Leeuw and Richard C. Sonnichsen, "Introduction: Evaluation and Organizational Learning: International Perspectives," in Frans L. Leeuw, Ray C. Rist, and Richard C. Sonnichsen, eds., *Can Governments Learn? Comparative Perspectives on Evaluation & Organizational Learning* (New Brunswick, NJ: Transaction Publishers, 1994), p. 1.

41. Christopher J. Bosso, "The Contextual Bases of Problem Definition," in David A. Rochefort and Roger W. Cobb, eds., *The Politics of Problem Definition* (Lawrence: University Press of Kansas, 1994), pp. 182–203.
42. Frans L. Leeuw and Richard C. Sonnichsen, "Introduction: Evaluation and Organizational Learning: International Perspectives," in Frans L. Leeuw, Ray C. Rist, and Richard C. Sonnichsen, eds., *Can Governments Learn? Comparative Perspectives on Evaluation & Organizational Learning* (New Brunswick, NJ: Transaction Publishers, 1994), pp. 1–2.
43. Robert T. Stout, Marilyn Tallerico, and Kent P. Scribner, "Values: The 'What?' of the Politics of Education," in Jay D. Scribner and Donald H. Layton, eds., *The Study of Educational Politics* (London: Falmer Press, 1995), pp. 5–20.
44. Mary E. Hawkesworth, *Theoretical Issues in Policy Analysis* (Albany: State University of New York Press, 1988), p. 4.
45. Pauline M. Rosenau, *Post-Modernism and the Social Sciences: Insights, Inroads, and Intrusions* (Princeton: Princeton University Press, 1992), p. 63.
46. Mary E. Hawkesworth, *Theoretical Issues in Policy Analysis* (Albany: State University of New York Press, 1988), p. 188.
47. Larry Cuban, "Why Do Some Reforms Persist?" *Educational Administration Quarterly* 24:3 (1988), p. 329.
48. Donald R. Warren, "Passage of Rites: On the History of Educational Reform in the United States," in Joseph Murphy, ed., *The Educational Reform Movement of the 1980s: Perspectives and Cases* (Berkeley, CA: McCutchan, 1990), pp. 57–81.
49. Larry Cuban, "Why Do Some Reforms Persist?" *Educational Administration Quarterly* 24:3 (1988), pp. 329–335.
50. Yehezkel Dror, *Policymaking Under Adversity* (New Brunswick: Transaction Books, 1986).
51. Ray C. Rist, "The Preconditions for Learning: Lessons from the Public Sector," in Frans L. Leeuw, Ray C. Rist, and Richard C. Sonnichsen, eds., *Can Governments Learn? Comparative Perspectives on Evaluation and Organizational Learning* (New Brunswick, NJ: Transaction Publishers, 1994), pp. 189–205.
52. Ray C. Rist, "The Preconditions for Learning: Lessons from the Public Sector," in Frans L. Leeuw, Ray C. Rist, and Richard C. Sonnichsen, eds., *Can Governments Learn? Comparative Perspectives on Evaluation and Organizational Learning* (New Brunswick, NJ: Transaction Publishers, 1994), p. 197.
53. Gary Mucciaroni, *The Political Failure of Employment Policy, 1945–1982* (Pittsburgh: University of Pittsburgh Press, 1990), p. 258.
54. Paul A. Sabatier and Hank C. Jenkins-Smith, eds., *Policy Change and Learning: An Advocacy Coalition Approach* (Boulder, CO: Westview Press, 1993).
55. David B. Tyack, *The One Best System* (Cambridge, MA: Harvard University Press, 1974).

56. Ray C. Rist, "The Preconditions for Learning: Lessons from the Public Sector," in Frans L. Leeuw, Ray C. Rist, and Richard C. Sonnichsen, eds., *Can Governments Learn? Comparative Perspectives on Evaluation and Organizational Learning* (New Brunswick, NJ: Transaction Publishers, 1994), pp. 189–205.
57. Ray C. Rist, "The Preconditions for Learning: Lessons from the Public Sector," in Frans L. Leeuw, Ray C. Rist, and Richard C. Sonnichsen, eds., *Can Governments Learn? Comparative Perspectives on Evaluation and Organizational Learning* (New Brunswick, NJ: Transaction Publishers, 1994), p. 197.
58. Colin J. Bennett and Michael Howlett, "The Lessons of Learning: Reconciling Theories of Policy Learning and Policy Change," *Policy Sciences* 25:3 (1992), p. 276.
59. Colin J. Bennett and Michael Howlett, "The Lessons of Learning: Reconciling Theories of Policy Learning and Policy Change," *Policy Sciences* 25:3 (1992), p. 290.
60. Bruce S. Cooper, Lance D. Fusarelli, and E. Vance Randall, *Better Policies, Better Schools: Theory and Application* (Boston: Allyn and Bacon, 2003).
61. While the Ohio and Wisconsin legislatures have passed pilot voucher plans in Cleveland and Milwaukee, respectively, the legislatures have resisted attempts to expand vouchers statewide; doing so requires an entirely different political calculus.
62. Sandra Vergari, "Conclusions," in Sandra Vergari, ed., *The Charter School Landscape* (Pittsburgh: University of Pittsburgh Press, 2002), p. 272.
63. Kerry A. White, "NRC Report Calls for Voucher Experiment," *Education Week* 1:2 (September 15, 1999), p. 3.
64. Jeff Archer, "Sanders 101," *Education Week* 18:34 (May 5, 1999), pp. 26–28.
65. Susan H. Fuhrman, "The Politics of Coherence," in Susan H. Fuhrman, ed., *Designing Coherent Education Policy* (San Francisco: Jossey-Bass, 1993), pp. 1–34.
66. James P. Spillane, J. P. and N. E. Jennings, "Aligned Instructional Policy and Ambitious Pedagogy: Exploring Instructional Reform from the Classroom Perspective," *Teachers College Record* 98:3 (1997), p. 450.
67. National Governors Association, 2002, "Maintaining Progress Through Systemic Education Reform," available at: http://www.nga.org/cda/files/000125EDREFORM.pdf (accessed June 16, 2002); James J. Scheurich, Linda Skrla, and Joseph J. Johnson, "State Accountability Policy Systems and Educational Equity," paper presented at the Annual Meeting of the American Educational Research Association, Seattle, WA, 2001.
68. Lance D. Fusarelli, "Tightly Coupled Policy in Loosely Coupled Systems: Institutional Capacity and Organizational Change," *Journal of Educational Administration* 40:6 (November 2002), pp. 561–575.

69. No agreement exists in the research literature on the effectiveness of various school choice proposals—including charter schools or vouchers. Furthermore, a number of scholars question the seemingly value-neutral assumptions of charter school accountability, asserting that charter school evaluation and charter renewal is not transparent and is as much a political process as a technical one. See Lance D. Fusarelli, "The Political Construction of Accountability," *Education and Urban Society* 33:2 (2001), pp. 157–169; Frederick M. Hess, "Whaddya Mean You Want to Close My School? The Politics of Regulatory Accountability in Charter Schooling" *Education and Urban Society* 33:2 (2001), pp. 141–156.
70. Hank C. Jenkins-Smith and Paul A. Sabatier, "The Dynamics of Policy-Oriented Learning," in Paul A. Sabatier and Hank C. Jenkins-Smith, eds., *Policy Change and Learning: An Advocacy Coalition Approach* (Boulder, CO: Westview Press, 1993), p. 52.
71. The Joint Select Committee concluded that "studies of school choice plans consistently find that educators, parents, and students are happier in schools they have chosen." Accordingly, the Committee recommended "experimentation with all types of public school choice and charter options" (Joint Select Committee, *Final Report of the Joint Select Committee to Review the Central Education Agency* (Austin: Texas Legislative Council, 1994), p. 25).
72. *ATPE Political Action Kit*, 1998; "Steady as She Goes," *Education Week* 16 (January 22, 1997), pp. 214–216.
73. CPS Position Paper, *The Choice Issue: Post Senate Bill 1 and Beyond* (Austin, no date).
74. A. Phillips Brooks, "Charter Schools to Open Amid Some Controversy," *Austin American-Statesman* (August 8, 1996), pp. A1, A12.
75. A. Phillips Brooks, "Charter Schools to Open Amid Some Controversy," *Austin American-Statesman* (August 8, 1996), pp. A1, A12.
76. Joint Select Committee, *Final Report of the Joint Select Committee to Review the Central Education Agency* (Austin: Texas Legislative Council, 1994).
77. A. Phillips Brooks, "Education to Remain a Priority, Bush Says," *Austin American-Statesman* (November 13, 1996), p. B3.
78. A. Phillips Brooks, "Texas Doing Homework on Charter Schools Concept," *Austin American-Statesman* (February 20, 1994), pp. A1, A17.
79. A. Phillips Brooks, "Charter Schools to Open Amid Some Controversy," *Austin American-Statesman* (August 8, 1996), pp. A1, A12.
80. J. Berls, "Parents Seek to Create Minority Charter Schools," *Austin American-Statesman* (March 21, 1996), p. B5.
81. TEPSA represents more than 4,900 school administrators.
82. H.B. 318 was authored by state representative Henry Cuellar (D) and sponsored in the Senate by Teel Bivins (R). In addition to expanding the number of open-enrollment charters to 100, H.B. 318 authorized an unlimited number of additional charters for at-risk students. H.B. 318

was signed into law by Governor Bush on June 17, 1997; Legislative Budget Board, *Fiscal Note* (May 31, 1997).

83. Source: www.heritage.org/schools/texas.html.
84. Lance D. Fusarelli, "Texas: Charter Schools and the Struggle for Equity," in Sandra Vergari, ed., *The Charter School Landscape* (Pittsburgh: University of Pittsburgh Press, 2002), pp. 175–191.
85. Although it is beyond the scope of this book to exhaustively examine the performance of charter schools, Sandra Vergari's edited book, *The Charter School Landscape* (Pittsburgh: University of Pittsburgh Press, 2002) examines charter school performance in 11 states and in Alberta, Canada; in another study, Jose da Costa and Frank Peters report that students attending the majority of charter schools in Alberta are performing "above the provincial average in all tested subjects at all grade levels" (Jose da Costa and Frank Peters, "Achievement in Alberta's Charter Schools," paper presented at the Annual Meeting of the University Council for Education Administration [November 1, 2002], p. 34).
86. *CER Newswire* 4:42 (October 15, 2002). Available at: www.edreform.com.
87. Lucy Hood, "Report Targets Teacher Shortage, Charter Schools" as cited in Lance D. Fusarelli, "Texas: Charter Schools and the Struggle for Equity," in Sandra Vergari, ed., *The Charter School Landscape* (Pittsburgh: University of Pittsburgh Press, 2002), p. 186.
88. "Charter Schools Worse on TAAS" as cited in Lance D. Fusarelli, "Texas: Charter Schools and the Struggle for Equity," in Sandra Vergari, ed., *The Charter School Landscape* (Pittsburgh: University of Pittsburgh Press, 2002), p. 184.
89. Shelley Kofler, "Charter School Report Highlights Problems," as cited in Lance D. Fusarelli, "Texas: Charter Schools and the Struggle for Equity," in Sandra Vergari, ed., *The Charter School Landscape* (Pittsburgh: University of Pittsburgh Press, 2002), p. 186.
90. "Charter Schools Worse on TAAS," *Austin American-Statesman* (December 19, 1999), p. A1.
91. Mark Vosburgh, "Group Gives Charter Schools an Incomplete," *The Plain Dealer* (March 30, 2000), p. B6.
92. Scott Stephens, "Charter Schools Don't Do Well on State Exams," *The Plain Dealer* (June 27, 2000), p. B1.
93. Scott Stephens, "Charter Schools Don't Do Well on State Exams," *The Plain Dealer* (June 27, 2000), p. B1.
94. Scott Stephens, "Charter Schools Don't Do Well on State Exams," *The Plain Dealer* (June 27, 2000), p. B1.
95. Scott Stephens, "Charter Schools Don't Do Well on State Exams," *The Plain Dealer* (June 27, 2000), p. B1.
96. Lynn Schnailberg, "Okla., Ore. Bump up Charter Law States to 36," *Education Week* 18:41 (June 23, 1999), p. 23.
97. Priscilla Wohlstetter, Noelle C. Griffin, and Derrick Chau, "Charter Schools in California: A Bruising Campaign for Public School Choice,"

in Sandra Vergari, ed., *The Charter School Landscape* (Pittsburgh: University of Pittsburgh Press, 2002), p. 52.

98. Bryan C. Hassel and Sandra Vergari, "Charter-Granting Agencies: The Challenges of Oversight in a Deregulated System," *Education and Urban Society* 31:4, 406–428.
99. Lance D. Fusarelli, "Texas: Charter Schools and the Struggle for Equity," in Sandra Vergari, ed., *The Charter School Landscape* (Pittsburgh: University of Pittsburgh Press, 2002), pp. 175–191.
100. Jay D. Scribner, Pedro Reyes, and Lance D. Fusarelli, "Education Politics and Policy: And the Game Goes On," in Jay D. Scribner and Donald H. Layton, eds., *The Study of Educational Politics* (Washington, D.C.: Falmer Press), pp. 201–212.
101. A. Phillips Brooks, "Choice Programs Won't Fix Public Schools, Study Says," *Austin American-Statesman* (March 30, 1994), pp. B1–B2.
102. A. Phillips Brooks, "Choice Programs Won't Fix Public Schools, Study Says," *Austin American-Statesman* (March 30, 1994), p. B1.
103. A. Phillips Brooks, "Choice Programs Won't Fix Public Schools, Study Says," *Austin American-Statesman* (March 30, 1994), p. B1.
104. A. Phillips Brooks, "Choice Programs Won't Fix Public Schools, Study Says," *Austin American-Statesman* (March 30, 1994), p. B2.
105. See E. Rasell and Richard Rothstein, eds., *School Choice: Examining the Evidence* (Washington, D.C.: Economic Policy Institute, 1993).
106. Nancy McClaran, "'Choice' is no Panacea for Schools," *Austin American-Statesman* (March 15, 1994), p. A11.
107. Senator Bivins has been labeled an enemy of public education by TSTA; A. Phillips Brooks, "Voucher Opponents Say They Can Block Bill in Senate," *Austin American-Statesman* (March 26, 1997), p. B9.
108. K. Shannon, "Bush Endorses Flexible Approach for Public Schools," *Austin American-Statesman* (July 2, 1996), p. B3.
109. A. Phillips Brooks, "Education to Remain a Priority, Bush Says," *Austin American-Statesman* (November 13, 1996), pp. B1, B3.
110. Press release, Office of State Senator Jeff Wentworth (April 6, 1999).
111. CSSB 1206 failed to make it to the floor of the Senate during the 1997 legislative session.
112. Press Release, Office of State Senator Teel Bivins (March 12, 1997).
113. Jessica L. Sandham, "Ohio Lawmakers Reinstate Voucher Program," *Education Week* (July 14, 1999), p. 17.
114. Mark Skertic and Michael Hawthorne, "Expansion in Ohio Unlikely Now," *Cincinnati Enquirer* (January 5, 1997), p. 1.
115. Mark Walsh, "Ohio Court Issues Mixed Verdict on Voucher Program," *Education Week* 18:38, p. 16.
116. Peter J. Shelly and Frank Reeves, "Common Ground Sought on Vouchers," *Pittsburgh Post-Gazette* (June 6, 1999), p. C5.
117. Peter J. Shelly and Frank Reeves, "Common Ground Sought on Vouchers," *Pittsburgh Post-Gazette* (June 6, 1999), p. C5.

118. Frank Reeves and Peter J. Shelly, "Compromise Bill Offers Vouchers to all PA. School Districts," *Pittsburgh Post-Gazette* (June 15, 1999), p. A1; Frank Reeves and Peter J. Shelly, "Lawmakers May Vote on Tuition Voucher Bill Today," *Pittsburgh Post-Gazette* (June 16, 1999), p. B1.
119. Alex Molnar, "Unfinished Business in Milwaukee," *Education Week* (November 17, 1999), p. 60.
120. Jeff Archer, "Positive Voucher Audit Still Raises Questions," *Education Week* 19:23 (February 16, 2000), p. 3.
121. Jeff Archer, "Positive Voucher Audit Still Raises Questions," *Education Week* 19:23 (February 16, 2000), p. 3.
122. Jeff Archer, "Positive Voucher Audit Still Raises Questions," *Education Week* 19:23 (February 16, 2000), p. 3.
123. H. Baldersheim, "Reforming Local Government Policymaking and Management Through Organizational Learning and Experimentation: The Case of Norway," *Policy Studies Journal* 21:1 (1993), pp. 104–114; Peter A. Hall, "Conclusion: The Political Power of Economic Ideas," in Peter A. Hall, ed., *The Political Power of Economic Ideas: Keynesianism Across Nations* (Princeton: Princeton University Press, 1989), pp. 361–392; Peter A. Hall, "Policy Paradigms, Social Learning, and the State: The Case of Economic Policymaking in Britain," *Comparative Politics* 25:3 (1993), pp. 275–296; Richard Rose, "What is Lesson-Drawing?" *Journal of Public Policy* 11 (1991), pp. 3–30; Richard Rose, *Lesson-Drawing in Public Policy* (Chatham, NJ: Chatham House, 1993).
124. This is not to minimize the extent of politics in schools. Politics sometimes plays a role in personnel decisions, particularly in hiring superintendents, and school politics has been known to affect decisionmaking.
125. Mary E. Hawkesworth, *Theoretical Issues in Policy Analysis* (Albany: State University of New York Press, 1988), p. 4.
126. Larry Cuban, "Why Do Some Reforms Persist?" *Educational Administration Quarterly* 24:3 (1988), p. 329.
127. D. Callahan and B. Jennings, "Introduction," in D. Callahan and B. Jennings, eds., *Ethics, the Social Sciences and Policy Analysis* (New York: Plenum Press, 1983), p. xiii.
128. Frank W. Lutz, "Policy-oriented Research: What Constitutes Good Proof," *Theory into Practice* 27:2 (1988), pp. 126–131; Deborah A. Stone, *Policy Paradox and Political Reason* (U.S.: Harper Collins, 1988); Carol H. Weiss, "Introduction," in Carol H. Weiss, ed., *Using Social Research in Public Policymaking* (Lexington, MA: D. C. Heath and Company, 1977), pp. 1–22.
129. Michael Mintrom, "Michigan's Charter School Movement: The Politics of Policy Design," in Sandra Vergari, ed., *The Charter School Landscape* (Pittsburgh: University of Pittsburgh Press, 2002), pp. 74–92; Sandra Vergari, "Charter Schools: A Primer on the Issues," *Education and Urban Society* 31:4, pp. 389–405.

130. Lance D. Fusarelli, "Texas: Charter Schools and the Struggle for Equity," in Sandra Vergari, ed., *The Charter School Landscape* (Pittsburgh: University of Pittsburgh Press, 2002), pp. 175–191.
131. Tracey Bailey, Carolyn Lavely, and Cathy Wooley-Brown, "Charter Schools in Florida: A Work in Progress," in Sandra Vergari, ed., *The Charter School Landscape* (Pittsburgh: University of Pittsburgh Press, 2002), p. 200.
132. Tracey Bailey, Carolyn Lavely, and Cathy Wooley-Brown, "Charter Schools in Florida: A Work in Progress," in Sandra Vergari, ed., *The Charter School Landscape* (Pittsburgh: University of Pittsburgh Press, 2002), p. 200.
133. Joseph P. Viteritti, "School Choice: Beyond the Numbers," *Education Week* 19:24 (February 23, 2000), p. 44.
134. Joseph P. Viteritti, "School Choice: Beyond the Numbers," *Education Week* 19:24 (February 23, 2000), p. 38.
135. Sandra Vergari, "Conclusions," in Sandra Vergari, ed., *The Charter School Landscape* (Pittsburgh: University of Pittsburgh Press, 2002), p. 272.
136. Lance D. Fusarelli, "The Political Construction of Accountability," *Education and Urban Society* 33:2 (2001), pp. 157–169; Frederick M. Hess, "Whaddya Mean You Want to Close My School? The Politics of Regulatory Accountability in Charter Schooling" *Education and Urban Society* 33:2 (2001), pp. 141–156.
137. David K. Cohen and Heather C. Hill, *Learning Policy: When State Education Reform Works* (New Haven: Yale University Press, 2001), p. 187.
138. David K. Cohen and Heather C. Hill, *Learning Policy: When State Education Reform Works* (New Haven: Yale University Press, 2001), p. 11.

Chapter 6 The Political Dynamics of School Choice: Lessons Learned

1. V. O. Key, *Southern Politics in State and Nation* (New York: Vintage Books, 1949).
2. Robert C. Bulman and David L. Kirp, "The Shifting Politics of School Choice," in Stephen D. Sugarman and Frank R. Kemerer, eds., *School Choice and Social Controversy* (Washington, D.C.: Brookings Institution Press, 1999), p. 37.
3. Kent Grusendorf, "School Choice Deserves a Chance," *Austin American-Statesman* (March 30, 1994), p. A13.
4. A. Phillips Brooks, "Delco: Texas Must Revitalize its Schools," *Austin American-Statesman* (March 24, 1994), p. B2.
5. Hubert Morken and Jo Renee Formicola, *The Politics of School Choice* (Lanham, MD: Rowman & Littlefield, 1999), p. 7.
6. Eric Hirsch, "Colorado Charter Schools: Becoming an Enduring Feature of the Reform Landscape," in Sandra Vergari, ed., *The Charter*

School Landscape (Pittsburgh: University of Pittsburgh Press, 2002), pp. 93–112.

7. Michael Mintrom, *Policy Entrepreneurs and School Choice* (Washington, D.C.: Georgetown University Press, 2000), p. 61.
8. Hubert Morken and Jo Renee Formicola, *The Politics of School Choice* (Lanham, MD: Rowman & Littlefield, 1999), p. 7.
9. Michael Mintrom and David N. Plank, "School Choice in Michigan," in Paul E. Peterson and David E. Campbell, eds., *Charters, Vouchers, and Public Education* (Washington, D.C.: Brookings Institution Press, 2001), pp. 43–58.
10. Michael Mintrom and David N. Plank, "School Choice in Michigan," in Paul E. Peterson and David E. Campbell, eds., *Charters, Vouchers, and Public Education* (Washington, D.C.: Brookings Institution Press, 2001), p. 55
11. Michael Mintrom and David N. Plank, "School Choice in Michigan," in Paul E. Peterson and David E. Campbell, eds., *Charters, Vouchers, and Public Education* (Washington, D.C.: Brookings Institution Press, 2001), p. 55.
12. Joseph P. Viteritti, *Choosing Equality: School Choice, the Constitution, and Civil Society* (Washington, D.C.: The Brookings Institution, 1999), p. 72.
13. Hubert Morken and Jo Renee Formicola, *The Politics of School Choice* (Lanham, MD: Rowman & Littlefield, 1999), p. 59.
14. Hubert Morken and Jo Renee Formicola, *The Politics of School Choice* (Lanham, MD: Rowman & Littlefield, 1999), p. 67.
15. Frederick M. Hess and Robert Maranto, "Letting a Thousand Flowers (and Weeds) Bloom: The Charter Story in Arizona," in Sandra Vergari, ed., *The Charter School Landscape* (Pittsburgh: University of Pittsburgh Press, 2002), p. 276.
16. Bryan C. Hassel, *The Charter School Challenge* (Washington, D.C.: Brookings Institution Press, 1999), p. 65.
17. Michael Mintrom and David N. Plank, "School Choice in Michigan," in Paul E. Peterson and David E. Campbell, eds., *Charters, Vouchers, and Public Education* (Washington, D.C.: Brookings Institution Press, 2001), p. 53.
18. Policy entrepreneurship models of policymaking were not discussed in this text, in large measure because all policy entrepreneurs must work within the domains of political culture, language, the institutional structure, and amongst competing interest groups and advocacy coalitions. Readers interested in the role of policy entrepreneurs should consult the following sources: Michael Mintrom and Sandra Vergari, "Advocacy Coalitions, Policy Entrepreneurs, and Policy Change," *Policy Studies Journal* 24:3 (1996), pp. 420–434; Michael Mintrom, *Policy Entrepreneurs and School Choice* (Washington, D.C.: Georgetown University Press, 2000); Hubert Morken and Jo Renee Formicola, *The Politics of School Choice* (Lanham, MD: Rowman & Littlefield, 1999).

19. Joseph P. Viteritti, *Choosing Equality: School Choice, the Constitution, and Civil Society* (Washington, D.C.: The Brookings Institution, 1999), p. 71.
20. Chester E. Finn, Jr., Bruno V. Manno, and Gregg Vanourek, "Charter Schools: Taking Stock," in Paul E. Peterson and David E. Campbell, eds., *Charters, Vouchers, and Public Education* (Washington, D.C.: Brookings Institution Press, 2001), pp. 19–42.
21. The UFT, representing New York City teachers, is an affiliate of the American Federation of Teachers.
22. Joetta L. Sack, "Ohio Charters Targeted in Election Politics," *Education Week* 22:3 (September 18, 2002), pp. 17, 20.
23. Lance D. Fusarelli, "Texas: Charter Schools and the Struggle for Equity," in Sandra Vergari, ed., *The Charter School Landscape* (Pittsburgh: University of Pittsburgh Press, 2002), p. 190.
24. Eric Hirsch, "Colorado Charter Schools: Becoming an Enduring Feature of the Reform Landscape," in Sandra Vergari, ed., *The Charter School Landscape* (Pittsburgh: University of Pittsburgh Press, 2002), pp. 93–112.
25. Eric Hirsch, "Colorado Charter Schools: Becoming an Enduring Feature of the Reform Landscape," in Sandra Vergari, ed., *The Charter School Landscape* (Pittsburgh: University of Pittsburgh Press, 2002), p. 97.
26. Sandra Vergari, "Introduction," in Sandra Vergari, ed., *The Charter School Landscape* (Pittsburgh: University of Pittsburgh Press, 2002), pp. 6–7.
27. Michael Mintrom, "Michigan's Charter School Movement: The Politics of Policy Design," in Sandra Vergari, ed., *The Charter School Landscape* (Pittsburgh: University of Pittsburgh Press, 2002), p. 77.
28. Jane H. Karper and William Lowe Boyd, "Interest Groups and the Changing Environment of State Educational Policymaking: Developments in Pennsylvania," *Educational Administration Quarterly* 24:1 (February 1988), p. 50.
29. Robert C. Bulman and David L. Kirp, "The Shifting Politics of School Choice," in Stephen D. Sugarman and Frank R. Kemerer, eds., *School Choice and Social Controversy* (Washington, D.C.: Brookings Institution Press, 1999), p. 61.
30. Robert C. Bulman and David L. Kirp, "The Shifting Politics of School Choice," in Stephen D. Sugarman and Frank R. Kemerer, eds., *School Choice and Social Controversy* (Washington, D.C.: Brookings Institution Press, 1999), p. 61.
31. Robert C. Bulman and David L. Kirp, "The Shifting Politics of School Choice," in Stephen D. Sugarman and Frank R. Kemerer, eds., *School Choice and Social Controversy* (Washington, D.C.: Brookings Institution Press, 1999), p. 62.
32. Mark Skertic and Michael Hawthorne, "Expansion in Ohio Unlikely Now," *Cincinnati Enquirer* (January 5, 1997), p. 1.
33. Hubert Morken and Jo Renee Formicola, *The Politics of School Choice* (Lanham, MD: Rowman & Littlefield, 1999).

34. Virginia Baxt and Liane Brouillette, "The State, the Lobbyists, and Special Education Policies in Schools: A Case Study of Decision Making in Texas," *Journal of School Leadership* 9, p. 154.
35. Hubert Morken and Jo Renee Formicola, *The Politics of School Choice* (Lanham, MD: Rowman & Littlefield, 1999).
36. Hubert Morken and Jo Renee Formicola, *The Politics of School Choice* (Lanham, MD: Rowman & Littlefield, 1999).
37. Lance D. Fusarelli and Bruce S. Cooper, "Why the NEA and AFT Sought to Merge—and Failed," *School Business Affairs* 65:4 (1999), pp. 33–38.
38. Joseph P. Viteritti, *Choosing Equality: School Choice, the Constitution, and Civil Society* (Washington, D.C.: The Brookings Institution, 1999), p. 71.
39. Joseph P. Viteritti, *Choosing Equality: School Choice, the Constitution, and Civil Society* (Washington, D.C.: The Brookings Institution, 1999).
40. Charles Mahtesian, "Teachers Union Taught a Lesson in Political Arena," *The Plain Dealer* (December 14, 1995), p. A28.
41. Charles Mahtesian, "Teachers Union Taught a Lesson in Political Arena," *The Plain Dealer* (December 14, 1995), p. A28.
42. Robert C. Bulman and David L. Kirp, "The Shifting Politics of School Choice," in Stephen D. Sugarman and Frank R. Kemerer, eds., *School Choice and Social Controversy* (Washington, D.C.: Brookings Institution Press, 1999), p. 50.
43. Robert C. Bulman and David L. Kirp, "The Shifting Politics of School Choice," in Stephen D. Sugarman and Frank R. Kemerer, eds., *School Choice and Social Controversy* (Washington, D.C.: Brookings Institution Press, 1999), pp. 36–67.
44. Mark Walsh, "Charting the New Landscape of School Choice," *Education Week* 21:42 (July 10, 2002), pp. 1, 18–21.
45. Paul E. Peterson and David E. Campbell, "Introduction: A New Direction in Public Education?" in Paul E. Peterson and David E. Campbell, eds., *Charters, Vouchers, and Public Education* (Washington, D.C.: Brookings Institution Press, 2001), pp. 1–17.
46. John Gehring, "Voucher Battles Head to State Capitals," *Education Week* 21:42 (July 10, 2002), pp. 1, 24, 25.
47. Robert C. Johnston, "Governors: State Finances Worst Since World War II," *Education Week* (December 4, 2002), p. 18.
48. Peggy Fikac, "Perry Tries New Legislative Approach," *San Antonio Express-News* (December 26, 2002), p. B3.
49. Lance D. Fusarelli, "The Political Economy of Gubernatorial Elections: Implications for Education Policy," *Educational Policy* 16:1, pp. 139–160.
50. Erik W. Robelen, "Few Choosing Public School Choice for this Fall," *Education Week* 21:43 (August 7, 2002), pp. 1, 38, 39.
51. Erik W. Robelen, "Few Choosing Public School Choice for this Fall," *Education Week* 21:43 (August 7, 2002), pp. 1, 38, 39.

52. Erik W. Robelen, "Few Choosing Public School Choice for this Fall," *Education Week* 21:43 (August 7, 2002), pp. 1, 38, 39.
53. Erik W. Robelen, "Few Choosing Public School Choice for this Fall," *Education Week* 21:43 (August 7, 2002), pp. 1, 38, 39.
54. Catherine Gewertz, "Miami-Dade will Launch Choice Plan," *Education Week* 22:10 (November 6, 2002), pp. 1, 11.

About the Author

Lance D. Fusarelli is assistant professor in the Division of Educational Leadership, Administration, and Policy in the Graduate School of Education, Fordham University. He received his Ph.D. in educational administration, with a specialization in education politics and policy, from the University of Texas at Austin, his M.A. in government from UT-Austin, and his B.A. in history and American studies from Case Western Reserve University. He co-edited *The Promises and Perils Facing Today's School Superintendent* (Scarecrow Press, 2002) and co-authored *Better Policies, Better Schools* (Allyn and Bacon, in press). His research interests include the politics of education at the state level, school leadership (the superintendency), and school choice. His doctoral dissertation, upon which this book is in part based, received the Outstanding Dissertation Award from the Politics of Education Association.

INDEX